Homeowner's Guide to Concrete and Masonry

Robert Scharff

Ideals Publishing Corp.
Milwaukee, Wisconsin

Table of Contents

Working with Concrete 3

Concrete Foundations and Floors 17

Concrete Block Masonry 27

Brick and Stone Masonry 37

Structural Walls of Brick, Concrete Block, and Stone 48

Driveways, Sidewalks, and Patios 62

Masonry Projects for the Home 72

Decorative and Special Finishes for Masonry Surfaces 81

Repairing Concrete and Masonry Surfaces 87

Index 96

ISBN 0-8249-6115-3

Published by Ideals Publishing Corporation
11315 Watertown Plank Road
Milwaukee, Wisconsin 53226

Editor, David Schansberg

Drawings by Donald W. Holohan, Robert Mull, and Eric Werner

Cover designed by David Schansberg. Materials courtesy of Elm Grove Ace Hardware.

Cover photo by Jerry Koser

SUCCESSFUL
HOME IMPROVEMENT SERIES

Bathroom Planning and Remodeling
Kitchen Planning and Remodeling
Space Saving Shelves and Built-ins
Finishing Off Additional Rooms
Finding and Fixing the Older Home
Money Saving Home Repair Guide
Homeowner's Guide to Tools
Homeowner's Guide to Electrical Wiring
Homeowner's Guide to Plumbing
Homeowner's Guide to Roofing and Siding
Homeowner's Guide to Fireplaces
Home Plans for the '80s
Planning and Building Home Additions
Homeowner's Guide to Concrete and Masonry
Homeowner's Guide to Landscaping
Homeowner's Guide to Swimming Pools
Homeowner's Guide to Fastening Anything
Planning and Building Vacation Homes
Homeowner's Guide to Floors and Staircases
Home Appliance Repair Guide
Homeowner's Guide to Wood Refinishing
Children's Rooms and Play Areas
Wallcoverings: Paneling, Painting, and Papering
Money Saving Natural Energy Systems
How to Build Your Own Home

Working with Concrete

Concrete is a synthetic construction material made by mixing four ingredients together in proper proportions: cement, fine aggregate (usually sand), coarse aggregate (usually gravel or crushed stone), and water. A mixture of cement, sand, and water, without coarse aggregate, is not concrete but mortar or grout.

The fine and coarse aggregate in a concrete mix are called inert ingredients; the cement and water are active ingredients. The inert ingredients and the cement are thoroughly mixed together first. When water is added, a chemical reaction (called hydration) between the water and the cement begins, causing the concrete to harden.

Remember, the hardening process is caused by hydration, not by a drying out of the mix. Concrete must be kept as moist as possible during the initial hydration process. Drying out would cause a drop in water content below the amount required for the chemical reaction to work properly.

Because concrete hardens and does not dry, it will harden just as well under water as it will in the air. Concrete can be formed into practically any shape with a variety of finishes, textures, and colors. Since it is composed only of inorganic materials, concrete is impervious to decay and fire and resists termites and rodents. Concrete is unaffected by the cold of the North or the heat of the South. Concrete's basic ingredients, portland cement and aggregates, are available nationwide at reasonable prices. These many desirable properties combine to give concrete durability, good appearance, and long-time economy that no other building material can offer.

Choosing Ingredients

The strength and durability of concrete is determined by the kinds of cement and aggregates used, as well as the amount of water added. The type of mix is determined by the quality of concrete needed for a particular job. A knowledge of concrete's several components is necessary in home improvement.

Types of Cement Portland cement is an extremely fine powder manufactured in a cement plant. Portland cement forms a paste when mixed with water. The quality of this paste determines the strength and durability of the finished concrete. Too much mixing water makes the paste thin and weak.

Different types of portland cement are manufactured to meet certain physical and chemical requirements for specific purposes. The American Society for Testing and Materials (ASTM) lists the five types as follows:

- Type I (normal portland cement; frequently marked with an N) is used for all general types of construction: patios, driveways, sidewalks, steps, foundations, and soil-cement mixtures. Most normal portland cement is gray. If white concrete is needed or desired, as in decorative work, specify "White Portland Cement Type I."
- Type II (modified portland cement) has a lower heat of hydration than Type I, and lower heat generated by the hydration of the cement improves resistance to sulfate attack. This type is seldom used around the home.
- Type III (high-early-strength portland cement) is used where high strengths are desired at very early periods. It is used when forms must be removed to put the concrete in service as quickly as possible and in cold weather construction to reduce the period of protection against low temperatures.
- Type IV (low-heat portland cement) is a special cement used where the amount and rate of heat generated must be kept to a minimum.
- Type V (sulfate-resistant portland cement) is a cement intended for use only in structures exposed to high alkali conditions.

Air-entrained portland cement is a special cement that has been developed to produce concrete that has a resistance to freeze/thaw action and scaling caused by chemicals applied for severe frost and ice removal. In cold climates and mild climates that have several freezing and thawing cycles each year, it should be used for all exterior concrete work.

Air entrainment has other advantages that are desirable for interior use. For example, tiny air bubbles act like ball bearings in the mix, increasing its workability, thus resulting in the need for less mixing water.

To create air-entrained concrete, chemicals called air-entraining agents are added to the mixing water. Building-materials suppliers sometimes carry air-entraining agents. Ready-mix plants stock them for their own use and may sell you a small quantity. The amount to be added to the mix depends upon the brand of air-entraining agent used. The result should be approximately 6 percent entrained air.

Another less tedious method of obtaining air-entrained concrete is to purchase a portland cement that contains an interground air-entraining agent. These cements—available in all five types—are identified as "air-entraining," or with the letter A after the type and can be bought from the same suppliers that sell regular portland cements.

Portland cement is packed in cloth or paper sacks. A 94-pound sack of cement amounts to about 1 cubic foot by loose volume.

Cement will retain its quality indefinitely if it does not come in contact with moisture. If it is allowed to absorb appreciable moisture in storage, it will set more slowly and its strength will be reduced. Sacked cement should be stored in an enclosed area that is as watertight and airtight as possible. The sacks should be stacked against each other to prevent air circulation between them, but they should not be stacked against outside walls.

Water Water for mixing should be clean and free of oil, acid, and other foreign substances. Drinking water generally can be used for making cement.

Aggregates Aggregates are divided into two sizes, fine and coarse.

Natural sand is the most common fine aggregate. When ordering, ask for "concrete sand." The sand should have particles ranging in size from ¼ inch down to dust-size particles. Mortar sand should not be used for making concrete since it contains only small particles.

Gravel and crushed stone are the most common coarse aggregates. Rounded pieces are better than long, sliver-like pieces. Particles range in size from ¼ inch to a maximum of 1½ inches.

The most economical concrete mix is obtained by using coarse aggregate, with the largest practical maximum size equal to one-quarter the thickness of the finished concrete. Accordingly, 4-inch slabs may use coarse aggregate with 1-inch maximum size, while slabs of 6-inch thickness may use 1½-inch maximum-size aggregate. Use a 1-inch maximum-size aggregate for steps.

Both fine and coarse aggregates must be clean and free of dirt, clay, silt, coal, or other organic matter such as leaves and roots. Foreign matter prevents the cement from properly binding the aggregate particles together, resulting in porous concrete with low strength and durability.

If you suspect that the sand contains too much extremely fine material conduct a silt test. Fill an ordinary quart canning jar or milk bottle to a depth of 2 inches with a representative sampling of the sand taken from at least five different locations in

Aggregate for use in concrete should contain a range of sizes. Shown here is 1½-inch maximum-size aggregate. Pieces vary from ¼ to 1½ inches.

the sand pile. Add clean water to the sand until the jar or bottle is about three-quarters full. Shake the container vigorously for about a minute. Use the last few shakes to level off the sand. Allow the container to stand for an hour. Any clay or silt will settle above the sand. If this layer is more than ³⁄₁₆-inch thick, the sand is not satisfactory.

Suitable concrete aggregates have a full range of sizes but no excess amount of any one size. Large aggregates fill out the bulk of the concrete, minimizing the use of the more expensive cement. Smaller aggregates fill the spaces between the larger ones. An even distribution of sizes produces the most economical and workable concrete.

Buy fine and coarse aggregates separately from a reputable building-materials supplier. If there is a ready-mix producer in your area, buy from him. He can offer a wide range of aggregate types and sizes.

Store aggregates on a clean, hard surface, but not directly on the ground. Besides wasting material, ground storage can cause contamination with mud and dirt. Cover aggregate piles to keep rain out. Do not use the bottom layer of an uncovered aggregate pile.

Proportions The strength of concrete depends mostly on the proportion of ingredients. The most accurate way to measure ingredients is by weight. However, measuring by volume is accurate enough for projects around the house and is much easier.

The most frequently used formulas for mixing cement, sand, and coarse aggregate are listed here.

1. The best all-around mix is 1 part cement, 2¼ parts sand, and 3 parts coarse aggregate. This mix is the most impervious to water and should be used outdoors for walks, patios, swimming pools, and so on. Because of its strength, the mix is also used in pouring driveways and floors.

2. For foundation walls, footings, and interior concrete work, which are not exposed to snow and rain, use a mixture of 1 part cement, 2¾ parts sand, and 4 parts coarse aggregate.

3. The weakest mixture used in construction is 1 part cement, 3 parts sand, and 5 parts coarse aggregate. Large foundations and footings are built with this mix.

All three mixtures require 6 gallons of water per cubic foot of cement.

Ordering Ingredients To determine how much cement, sand, aggregate, and water you need for a project, first calculate the total amount of concrete required using the following formula.

$$\frac{\text{Width (ft.)} \times \text{Length (ft.)} \times \text{thickness (in.)}}{12} = \text{cubic feet}$$

For example, a 3-foot wide sidewalk that is 36 feet long and 4 inches thick would require 36 cubic feet of concrete. Add another 10 percent for loss of materials due to spills, uneven subgrade, mixing mistakes, and so on. This additional 10 percent would bring the total amount to 40 cubic feet of concrete. Since most ingredients are ordered in cubic yards, divide the total amount by 27. Forty cubic feet equals approximately 1.5 cubic yards of concrete.

Refer to the following table which lists proportions of sand, cement, aggregate, and water that produce 1 cubic yard of concrete. Multiply the number of cubic yards needed for the project by each quantity listed. Assuming that the sidewalk in the example would require 1:2¼:3 concrete, the order would be:

cement	1.5 × 6 bags = 9 bags
sand	1.5 × 0.51 cubic yard = 0.765 or 0.8 cubic yard
coarse aggregate	1.5 × 0.66 cubic yard = 0.99 or 1 cubic yard
water	1.5 × 36 gallons = 54 gallons

Materials to Make 1 Cubic Yard of Concrete	
1:2¼:3 concrete	
Materials	**Quantity**
cement	6 bags
sand (wet)	14 cubic feet or 0.51 cubic yard
coarse aggregate	18 cubic feet or 0.66 cubic yard
water	36 gallons
1:2¾:4 concrete	
Materials	**Quantity**
cement	5 bags
sand (wet)	14 cubic feet or 0.51 cubic yard
coarse aggregate	20 cubic feet or 0.74 cubic yard
water	30 gallons
1:3:5 concrete	
Materials	**Quantity**
cement	4.5 bags
sand (wet)	13 cubic feet or 0.48 cubic yard
coarse aggregate	22 cubic feet or 0.8 cubic yard
water	27 gallons

Note: It is not necessary to measure the ingredients to the exact amounts given. You can round off the quantities. The change will be negligible.

Measuring and Mixing Concrete Proper measuring is essential to making quality concrete. Once measured, ingredients must be thoroughly mixed so that cement paste coats every particle of fine and coarse aggregate in the mix.

It is best for the beginning concrete mason to use some type of measuring device like a large bucket, tub, or similar container and dump in one bag of cement. Mark the top of the cement on the side of the bucket to indicate 1 cubic foot volume. The marked bucket can be used to measure the other ingredients.

If you mix the same amount of concrete each time, make batching containers by measuring the correct amount of sand, cement, and aggregate into buckets. Mark the level in each bucket. Each time you mix a batch, fill the containers to the premarked levels. To measure water, use a bucket with known capacity.

Dry sand is rarely available for concrete work. Sand used on most jobs contains some moisture which must be accounted for as part of the mixing water. Check for water content of sand by squeezing the sand with your hand. Wet sand forms a ball and leaves no noticeable moisture on the palm. Damp sand falls apart when squeezed. Very wet sand forms a ball when you squeeze it and leaves moisture on your palm.

If possible, use wet sand. Avoid excess moisture by covering the pile with plastic. If you use damp or very wet sand and the trial batches of concrete you have mixed with the sand are too thin or too stiff, adjust the sand and water quantities slightly.

Machine Mixing The best way to mix concrete is with a concrete mixer. It ensures thorough mixing and is the only way to produce air-entrained concrete.

Small mixers from ½- to 6-cubic foot capacity can be rented or purchased. For extensive work around the home, it might pay to purchase a mixer. For the occasional small job, however, rent a mixer from your local rental service store or yard.

Mixers are powered by gasoline or electricity. The gasoline-powered mixer is more versatile because it can be operated anywhere. The electric-powered mixer is quieter and easier to operate, but it requires a convenient outlet.

Mixer sizes are designated according to the maximum concrete batch in cubic feet that can be mixed efficiently. This is approximately 60 percent of the total volume of the mixer drum. The maximum batch size is usually shown on an identification plate attached to the mixer. Never load a mixer beyond its maximum batch capacity. The choice of mixer size will depend upon the extent of your project and the amount of concrete you can handle in any one batch. To mix a 1-cubic foot batch of concrete you will have to handle 140 to 150 pounds of material.

For best results, load the ingredients into the mixer in the following sequence:

1. With the mixer stopped, add all the coarse aggregate and half of the mixing water. If an air-entraining agent is used, mix it with this part of the mixing water.
2. Start the mixer, then add the sand, cement, and remaining water.

After all of the ingredients are in the mixer, continue mixing for at least three minutes or until all materials are thoroughly mixed and the concrete has a uniform color.

Concrete should be poured in the forms as soon as possible after mixing. If the concrete shows signs of stiffening, remix it for about two minutes to restore its workability. If, after remixing, the concrete is still too stiff to be workable, discard it. Never add water to concrete that has stiffened to the point where remixing will not restore its workability.

Mix a trial batch of concrete using the proper proportions. Discharge a sample of concrete from the mixer into a wheelbarrow or onto a slab and check for stiffness and workability. If the sample is a smooth, plastic, workable mass that will place and finish well, the proportions were correct and need no adjustment. Work the concrete with a shovel and smooth it with a float or trowel. A good, workable mix should look like the sample shown in the illustration. The concrete should be just wet enough to stick together without crumbling. It should slide down, not run off, a shovel. A good workable mix has sufficient cement paste to completely fill the spaces between aggregate. Aggregate should not separate when the concrete is transported and placed in the forms. There should be sufficient sand-cement paste to give clean, smooth surfaces free from rough spots.

If the trial batch is too wet, too stiff, too sandy, or too stony, it will be necessary to adjust the proportions of aggregates used in the mix.

If the mix is too wet, it contains too little aggregate for the amount of cement paste. Add more sand and coarse aggregate, depending upon the wetness of the mix. Add them to the trial batch in the mixer and mix for at least one minute. If the mix is still too wet, add more sand and coarse aggregate until the desired consistency is obtained. Record the total volume of added sand and coarse aggregate. In subsequent batches, use the original quantity of aggregate but reduce the amount of water.

A   B   C

Testing for water content of sand: (A) Wet sand, ideal for concrete making, forms a ball but leaves no moisture when squeezed; (B) damp sand will fall apart; and (C) very wet sand will form a ball and leave moisture in your palm.

In a workable mix, all the spaces between coarse aggregates are filled with sand and cement paste. Ingredient proportions are correct.

If the mix is too stiff, it contains too much aggregate. In the next batch, reduce the amounts of sand and coarse aggregate. Indicate the new volumes of sand and coarse aggregate on the respective cans according to the adjustments. To save a stiff trial batch, cement and water may be added in the proportions of 1 part water to 2 parts cement. This will increase the amount of cement paste and make the concrete more workable. Never add water alone to a mix that is too stiff.

If the mix is too sandy, decrease the amount of sand and add an equivalent amount of coarse aggregate. Record the new volumes of sand and coarse aggregate, and correct the marks in the batch cans.

If the mix is too stony, decrease the amount of coarse aggregate and add an equivalent amount of sand. Record the new volumes of sand and coarse aggregate, and correct the marks in the batch cans.

Your adjusted trial batch proportions are your final mix proportions and need not be changed again for future batches as long as your sand and coarse aggregate remain the same.

Thoroughly clean the mixer after you finish using it before the concrete can harden. To clean the inside of the mixer drum, add water and a few shovels of coarse aggregate while the drum is turning. Follow this by hosing the drum with water. Vinegar will remove the thin cement films on the exterior. If concrete builds up inside the drum, scrape out the

This mix is too stiff and would be difficult to finish and place properly. It contains too much sand and coarse aggregate.

large chunks, then scrub it with a wire brush. Do not use heavy hammers or chisels that might damage the drum and blades. Remove stubborn buildup with a solution of 1 part hydrochloric acid (muriatic acid) in 3 parts of water. Allow 30 minutes for penetration, then scrape or brush the buildup and rinse with clear water. Warning: Hydrochloric acid is hazardous and toxic. Avoid skin contact and fumes. Wear rubber or plastic gloves and chemical safety goggles. If the acid is used indoors, provide adequate ventilation.

Dry the mixer drum thoroughly to prevent rusting, and store the mixer with the opening of the drum pointing down.

Hand Mixing For small jobs requiring a volume of concrete less than a few cubic feet, it is sometimes more convenient and less expensive to mix by hand.

Hand mixing is not vigorous enough to make air-entrained concrete, even if you use air-entraining cement or an air-entraining agent. Hand mixing, therefore, should not be used for concrete exposed to freezing/thawing conditions or deicers.

Hand mixing should be done on a clean, hard surface or in a wheelbarrow. Mud and dirt will contaminate and weaken the concrete. Spread out a measured quantity of sand. Then, pour the required amount of cement on the sand. Mix the cement and sand thoroughly by turning with a short-handled,

This mix is too wet. Concrete that looks like this is less durable and more likely to crack. It contains too little sand and coarse aggregate for the amount of cement paste.

(A) This mix contains too much sand and not enough coarse aggregate. It would be easy to place and finish, but likely to crack. (B) This mix contains too much coarse aggregate and not enough sand. It would be difficult to work with and the finished concrete would be honeycombed and porous.

Mixing concrete by hand.

square-end shovel until you have a uniform color. Evenly spread this mixture and dump the required quantity of coarse aggregate in a layer on top. The materials are again turned by shovel until the coarse aggregate has been uniformly blended with the mixture of sand and cement. Form a depression or hollow in the center of the pile and slowly add the water. Turn all the materials in toward the center and continue mixing until the water, cement, sand, and coarse aggregate have all been thoroughly mixed.

Premixed Concrete

Small jobs can usually be done with convenient prepackaged concrete mixes. Building-materials suppliers, home centers, and hardware stores sell prepackaged concrete mixes.

Two things are common to all premixes: all parts of the mix are correctly proportioned, and all contain cement, which acts as a binder when water is added. The content and, most importantly, the function of premixes can vary greatly.

1. Sand mix contains cement and sand and is good for repairs and projects where up to a 2-inch thickness of mix is sufficient.
2. Concrete mix (also called gravel mix) contains cement, sand, and gravel and is for use where extra strength is required.
3. Mortar mix is a mixture of sand, cement, and lime. Its distinguishing characteristic is its plasticity and easy workability. It is most useful for joints between bricks, stones, cinder blocks, and concrete blocks.
4. Waterproof mix is used as a top coat on surfaces subject to a lot of water. Use it to repair a below-grade crack in a foundation to prevent moisture penetration. Waterproof premixes contain cement, sand, and a waterproofing compound, such as polyvinyl acetate. If you cannot buy the premix, you can easily add waterproofing compound to a bag of sand mix.

Packages are available in different weights, but the most common sizes are 45 and 90 pounds. A 90-pound package usually makes $\frac{2}{3}$ cubic foot of concrete. All you do is add water. Directions for mixing and adding water are given on the bag.

Prepackaged mixes are most convenient for very small jobs requiring only a few cubic feet of concrete. For larger jobs up to 1 cubic yard (27 cubic feet), compare the cost of using prepackaged mixes with the cost of buying the separate ingredients.

Ready-Mixed Concrete

The most convenient and economical source of 3 cubic yards or more of concrete is a ready-mix producer. Producers can supply concrete to meet the requirements of any project. Ready-mixed concrete is sold by the cubic yard (27 cubic feet), and a producer will usually deliver any quantity greater than 1 cubic yard. Whether the project requires 1 cubic yard or many, the job of proportioning, weighing,

Ready-mix concrete is more convenient for large jobs.

mixing, and hauling will be done according to careful specifications to ensure concrete of uniform quality.

The cost of ready-mixed concrete varies with the distance hauled, size of order, day of delivery, unloading time, and type of mix. Call several reputable producers to compare prices.

Estimating the amount of concrete needed can be done using the following table and example.

	Estimating Cubic Yards of Concrete for Slabs*					
	Area in square feet (width × length)					
Thickness, inches	10	25	50	100	200	300
3	0.09	0.27	0.47	0.99	1.96	2.85
4	0.12	0.31	0.62	1.23	2.47	3.70
5	0.15	0.39	0.77	1.54	3.09	4.63
6	0.19	0.46	0.93	1.85	3.70	5.56

Does not allow for losses due to uneven subgrade, spillage, and other factors. Add 5 to 10 percent for such contingencies.

To find the amount of concrete required for a 4-inch thick driveway 11 feet wide by 41 feet long, first figure the number of square feet by multiplying 11 feet by 41 feet which gives 451 square feet. This can be rounded off to 450.

From the table: 300 square feet = 3.70 cubic yards
100 square feet = 1.23 cubic yards
50 square feet = 0.62 cubic yards
Total: 450 square feet = 5.55 cubic yards

With a perfect subgrade and no spillage, 5½ cubic yards might be enough. But, to be safe, order 6 cubic yards. It is better to have some concrete left over than to run short.

The following points should be included in the specifications for a concrete order:

1. Maximum-size aggregate (gravel, crushed stone) should not exceed one-quarter the slab thickness.
2. Minimum cement content should be no less than the amount specified for the particular maximum-size aggregate used. These cement contents are essential for proper finishing and strength development. In areas exposed to a number of freeze/thaw cycles or to the use of deicers, it is advisable to use a minimum cement content of 560 pounds per cubic yard.
3. Maximum slump should not exceed 50 percent. This slump mix will give a good, workable material. Stiffer mixes are harder to place and finish by hand. Very wet, soupy mixes will not make durable concrete.

    You can make your own slump test by putting a few holes in the bottom of a 2-pound, 8-inch high coffee can and filling it with the concrete mix. Turn the can over on a smooth, flat surface and lift the can clear. The amount that the concrete slumps is an indication of the mix's water content. If it drops more than 50 percent of its original height (the height of the can), the mix has too much water.

4. Compressive strength at 28 days should be no less than 3,500 pounds per square inch.

5. Air content is required to obtain good durability in all concrete that is exposed to freezing, thawing, or deicing salts. When this protection is not required, reduce the air content by one-third to reduce bleeding and segregation, and to improve workability and finishability.

In addition to the amount required, the producer will need to know where and when to deliver the concrete. Place the order at least a day ahead of time.

Preparing for the Concrete

Before the concrete is mixed or delivered, the subgrade must be prepared and forms must be provided. The method and amount of preparation for the subgrade depends on the project. Subgrade preparation for various projects are explained later.

The finished shape of the concrete will conform exactly to the shape of the form. Forms range from simple to elaborate, but forms for any concrete project must be tight, rigid, and strong. If forms are not tight, there will be a loss of concrete which may result in honeycomb or a loss of water that causes sand streaking. The forms must be braced and strong enough to hold the concrete. Special care should be taken in bracing and tying down forms in which the mass of concrete will be large at the bottom and taper toward the top. The concrete tends to lift the form above its proper elevation. If the forms are to be used again, they must be easily removed and re-erected without damage.

Forms for concrete work are generally constructed from three different materials:

Guide for Ordering Ready-Mix Concrete for Drives, Walks and Patios				
Maximum-Size Aggregate (in.)	Minimum Cement Content (lbs. per cu. yd.)	Maximum Slump (in.)	Compressive Strength at 28 days (lbs. per sq. in.)	Air Content (percent by volume)
⅜	610	4	3,500	7½ ± 1
½	590	4	3,500	7½ ± 1
¾	540	4	3,500	6 ± 1
1	520	4	3,500	6 ± 1
1½	470	4	3,500	5 ± 1

1. Earth. Earth forms are used in subsurface construction where the soil is stable enough to retain the desired shape of the structure. This type of form requires less excavation, and there is better settling resistance. Earth forms are generally used for footings and foundations.

2. Metal. Metal forms are used where added strength is required or where the construction will be duplicated at another location. Metal forms are seldom used on home-type projects because of their cost.

3. Wood. Wooden forms are the most common type used in building construction. They are economical, easy to handle and produce, and adaptable to many shapes. Reusing form lumber later for roofing, bracing, or similar purposes results in further economies. Lumber should be straight, strong, and only partially seasoned. Kiln-dried timber may swell when soaked with water from the concrete. If the boards are tight jointed, swelling causes bulging and distortion. If green lumber is used, allow for shrinkage or keep the forms wet until the concrete is in place. Softwoods such as pine, fir, and spruce make the best and most economical form lumber since they are light, easy to work with, and available in most regions. Lumber that comes in contact with the concrete should be surfaced on one side and on both edges. The surfaced side faces the concrete. The edges may be square, shiplap, or tongue and groove. The latter makes a more watertight joint and tends to prevent warping. Plywood is economical for wall and floor forms if it is made with waterproof glue. Plywood is more warp-resistant and can be reused more often than lumber. Plywood is available in a variety of thicknesses. The standard length is 8 feet. The ⅝- and ¾-inch thicknesses are most economical. Thinner sections will require more solid backing.

Pouring the Concrete

Once the subgrade is properly prepared and the forms are in place, you are ready to pour the concrete and then smooth and finish it off. Before beginning to pour the concrete, have the proper tools readily available. Tools you will need include those in the illustration plus a wheelbarrow, a hose, a 2 by 4 straightedge long enough to span a section of concrete to be poured, and a flat-edged shovel.

Placing Concrete If you are mixing by hand or using a small mixer, you can place and work the concrete by yourself, at your own speed. With a ready-mix truck, you may need help in placing the concrete.

Plan to mix the concrete near the site. If you are using ready-mix, plan to get the truck as close to the

Attach forms to stakes by nailing through the stake into the form. Use double-headed nails.

site as possible without driving over existing sidewalks or driveways. If the job site is hard to reach, ask the producer for suggestions before delivery.

Uniformly place the concrete to the full depth of the forms and as near as possible to final position. Start in a corner and do not drag or flow the concrete excessively. Overworking the mix causes water and fine material to come to the surface and aggregates to settle to the bottom, weakening the concrete. Spade concrete along the forms to compact it firmly and fill in all of the spaces.

Spreading and spading are best done with a short-handled, square-end shovel. Spreading can also be done with special concrete rakes or come-alongs, which are hoelike tools. Rakes and hoes can cause segregation (a separation of large pieces of gravel or stone from the mortar in the mix).

After the concrete has been spread and compacted to fill the forms, strikeoff and bull-floating, or darbying, should begin immediately. If water collects on the surface, the concrete is hardening too fast. Concrete should not be spread over too large an area before strikeoff, nor should a large area be struck off and allowed to remain before bull-floating or darbying.

Divide a large project into manageable sections. Key the sections to each other by nailing a board or metal strip on the inside of the form. When the con-

crete has hardened, remove the form. The mixture for the adjacent slab will flow into and lock into the key you formed.

Strikeoff Strikeoff is the removal of excess concrete. The tool used is a straightedge or strikeoff. Mechanical straightedges are made of metal or wood, and they are frequently equipped with a vibrator that compacts the concrete during strikeoff. For small jobs, a straight 2x4 is fine.

Concrete is struck off by moving the straightedge back and forth in a sawlike motion. A small roll of concrete should be kept ahead of the straightedge to fill in low spots. The straightedge should be tilted in the direction of travel to obtain a cutting edge. A second pass should be made, if needed, to remove any remaining bumps or low spots. The straightedge should be tilted in the opposite direction for the second pass.

Bull-floating or Darbying Bull-floating or darbying immediately follows the strikeoff to level ridges and fill voids left by the straightedge, as well as to embed coarse aggregate slightly below the surface. Use a bull float for areas too large to reach with a darby.

Bull floats are large, long-handled floats made of wood. Push the bull float with the front (toe) of the float raised so that it will not dig into the concrete surface. Then, pull it back with the float blade flat on the surface to cut off bumps and fill holes. If holes or depressions remain and no excess concrete is left on the slab, shovel additional concrete from a wheelbarrow and bull-float the surface again.

Darbies are made of 1x3s and 1x4s. Hold the darby flat against the surface of the concrete and work from right to left with a sawing motion, cutting off bumps and filling depressions. When the surface is level, the darby should be tilted slightly and again

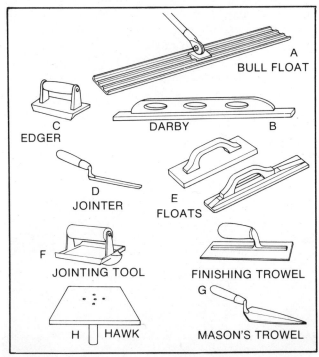

Tools for working concrete: bull float (a), darby (b), edger (c), jointer (d), floats (e), jointing tool (f), trowels (g), and hawk (h).

Spreading concrete with a square-end shovel.

moved from right to left to fill any small holes left by the sawing motion.

This operation should level, shape, and smooth the surface and work up a slight amount of cement paste. Overworking results in a less durable surface.

Finishing Concrete The finishing operations of edging, jointing, floating, and troweling should not be started until the concrete stiffens slightly, which depends upon the wind, temperature, and relative humidity. On hot, dry, windy days, the waiting period is very short. On cool, humid days, it can be several hours. With air-entrained concrete, there may be little or no waiting. Begin finishing air-entrained concrete before the surface gets dry and tacky.

Edging Edging is the first operation. It produces a neat, rounded edge that prevents chipping and compacts and hardens the surface next to the form where floats and trowels are less effective.

Edging tools are made of steel and bronze. Stainless-steel edgers with a ½-inch radius are recommended for walks, drives, and patios. Preliminary edging should be done with a longer and wider edging tool (approximately 10 x 6 inches).

Immediately after using the bull float or darby, cut the concrete away from the forms to a depth of 1 inch, using a pointed mason trowel. The concrete should have set sufficiently to hold the shape of the edger tool. The edger should be held flat on the concrete surface. Tilt the front of the edger up slightly when moving the tool forward. When moving the tool back over the edge, tilt the rear slightly. Be careful to prevent the edger from leaving too deep an impression, as these indentations may be difficult to remove later. In some cases, edging is required

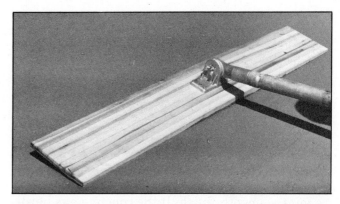

Correct blade position for pushing a bull float across a slab.

after each finishing operation. Final edging is done after final troweling.

Jointing Immediately after edging is done, control joints can be cut in the concrete. Proper jointing prevents random cracking.

Like any building material, concrete contracts and expands slightly with moisture and temperature changes. When newly placed, concrete attains its largest volume. When dry and cold, it contracts to its smallest volume. Unless provisions are made to control these natural changes, cracks may result. Effective control is obtained with control joints and isolation joints.

Control joints can be made with a hand tool or a circular saw with a silicon-carbide blade, or they can be formed by using wood divider strips. The tooled or sawed joint should extend into the slab one-fifth to one-fourth of the slab thickness. A cut of this depth provides a weakened section that induces cracking beneath the joint where it is not visible.

Striking off a patio section with a 2 by 4 straightedge.

Working the darby with a sawing motion.

Mark the location of each joint with a string or chalk line on both side forms and on the concrete surface. A straight 1-inch board at least 6 inches wide should be used to guide the jointing tool. Rest the board on the side forms perpendicular to the edges of the slab. Hold the jointing tool against the board as you move it across the slab.

To start the joint, push the jointing tool into the concrete and move it forward, applying pressure to the back of the tool. After the joint is cut, turn the tool around and pull it back over the groove to give the joint a smooth finish. If the concrete has hardened to the point where the jointing tool will not penetrate easily, a hand axe can be used to push it through the concrete.

Instead of being hand-tooled, control joints can be cut with a circular saw equipped with a masonry cutting blade. First score the line, then lower the blade slightly and make a second pass. Continue to lower the blade slightly and cut concrete in repeated passes until full joint depth is obtained. To function properly, a sawed joint must be cut as deep as a hand-tooled joint, one-fourth to one-fifth of the slab thickness. Sawing should be done 4 to 12 hours after the concrete hardens.

Edging and jointing must be done carefully. If the surface is gouged out by hand edgers and groovers, the marks may be difficult to remove later.

Floating Following edging and jointing, the surface is floated to: (1) embed large aggregate just below the surface; (2) remove surface imperfections; and (3) compact the concrete and consolidate mortar at the surface.

Hand floats are made of metal, wood, plastic, and composition materials. Magnesium floats are light and strong, and they slide easily over a concrete surface. Wood floats drag more, hence they require greater effort to use. If you want a rough-textured finish, use a wood float for the final finish.

Edging a concrete slab.

The hand float should be held flat on the concrete surface and moved with a slight sawing motion in a sweeping arc to fill in holes, cut off lumps, and smooth ridges.

Floating produces a fairly even, but not smooth, texture. Since this texture has good skid-resistance, floating is often used as a final finish. It may be necessary to float the surface a second time, after some hardening has taken place, to impart the desired final texture to the concrete. If edges or control joints are marred by floating, repeat edging and jointing after each floating.

Troweling Troweling produces a smooth, hard, dense surface and should never be done on a surface that has not been floated.

Cutting concrete away from the form before edging.

Using a jointing tool to cut control joints.

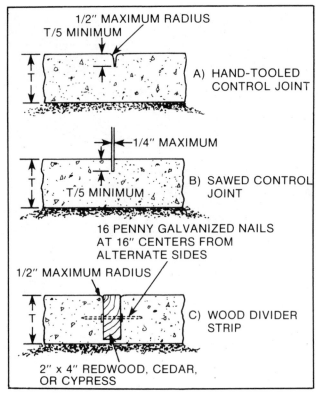

Three types of control joints.

Jointing tools or groovers, like edgers, are made of stainless steel and other metals and are available in various sizes and styles. The radius of a jointing tool should be ¼ to ½ inch. Jointing tools with worn-out or shallow bits should not be used.

On large slabs where you cannot reach the entire surface, kneel on a board placed on the slab. Float and immediately trowel an area before moving the board. Timing is important. The concrete must be hard enough so that water and fine material are not brought to the surface, but not too hard to finish. The tendency in many cases is to start when the concrete is too soft. Premature finishing may cause scaling or dusting.

Hand trowels are made of high-quality steel in various sizes. For the first troweling, use a larger

Cutting a control joint with a circular saw and an abrasive blade.

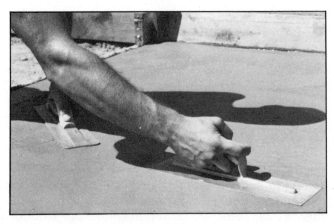

Floating with a magnesium float.

tool, 18 by 42¾ inches for example. Shorter and narrower trowels are used for additional trowelings as the concrete sets and becomes harder. A 12- by 3-inch trowel, known as a fanning trowel, is recommended for the final troweling. When one trowel is used for the entire operation, choose one that is 14 by 4 inches. Power trowels can be rented for major projects.

For the first troweling, the trowel blade should be held flat on the surface. If it is tilted, permanent ripples can form. The hand trowel is used in a sweeping arc motion, each pass overlapping one-half of the previous pass. The first troweling may produce a good, smooth surface, but additional trowelings will increase smoothness and hardness.

Between trowelings, permit the concrete to become slightly harder. Use a smaller trowel with the blade tilted slightly for the second troweling.

If the desired finish is not obtained with the second troweling, a third troweling should follow after the surface is allowed to harden more. The final pass should make a ringing sound as the tilted blade moves over the hardening surface.

If necessary, tooled edges and joints should be rerun after troweling.

Troweling with a steel hand trowel.

Other Finishes A smooth-troweled slab is easy to clean but can be slippery when wet. For better footing on sidewalks and outdoor patios, the slab can be roughened slightly by brooming or by using one of the textured special finishes described later.

Curing Concrete

Curing is one of the most important steps in concrete construction. Proper curing increases the strength and durability of concrete. The hardening of concrete which happens during curing is brought about by the chemical reaction between cement and water, called hydration. Proper water content and temperature must be maintained. In near freezing temperatures hydration slows greatly. When it is too hot, dry, or windy, water is lost by evaporation and hydration stops. Ideal conditions for hardening are moist and warm.

Moist-curing is done in a number of ways: by covering the surface with wet burlap, by keeping the surface wet with a lawn sprinkler, or by sealing the concrete surface with plastic sheeting, waterproof paper, or curing compound to prevent moisture loss.

If burlap is used, it should be free of chemicals that could weaken or discolor the concrete. New burlap should be washed before use. Place it when the concrete is hard enough to withstand surface damage and sprinkle it periodically to keep the concrete surface continuously moist.

Water curing with lawn sprinklers, nozzles, or soaking hoses must be continuous to prevent interruption of the curing process.

Wet-curing with burlap.

Curing with plastic sheets is convenient. They must be laid flat, thoroughly sealed at joints, and anchored carefully along edges. Curing with plastic can cause patchy discoloration in colored concrete. Chemical curing compounds are better for colored concrete.

Pigmented curing compounds provide the easiest and most convenient method of curing. These compounds are applied by spraying soon after the final finishing operation. The surface should be damp, but not wet. Complete coverage is essential.

Use of curing compounds is not recommended during late fall in northern climates where deicers are used to melt ice and snow. Using curing compounds at that time may prevent proper air-drying of the concrete, which is necessary to enhance its resistance to damage caused by deicers.

Curing should be started as soon as possible and should continue for a period of five days in warm

Applying a broom finish.

Curing with plastic.

weather (70 degrees Fahrenheit or higher) or seven days in cooler weather (50 to 70 degrees Fahrenheit). The temperature of the concrete must not be allowed to fall below 50 degrees Fahrenheit during the curing period.

Curing During Cold Weather Ideally, walks, driveways, patios, and steps should be built well in advance of cold weather. When placed during warm weather, there is sufficient time for concrete to develop strength to resist freezing, thawing, and chemical deicers. However, a good job can be done during cold weather if proper methods are carefully followed. Heated concrete should be ordered from a ready-mix company so that the temperature of the mix does not fall below 50 degrees Fahrenheit during placing, finishing, and curing. When there is danger of freezing, concrete should be kept warm during curing. Insulating blankets or 12 to 24 inches of dry straw covered with canvas, waterproof paper, or plastic sheeting to keep it dry and in place can be used. The effectiveness of the protection can be checked by placing a thermometer under the covering. Slab edges and corners are most vulnerable to freezing.

During cold weather, high-early-strength concrete may be used to speed up setting time and strength development. This can reduce the curing period, but a minimum temperature of 50 degrees Fahrenheit must be maintained. High-early-strength concrete is made by using Type III cement or an extra 100 pounds of cement in each cubic yard of concrete.

Curing in Hot Weather In hot weather, precautions may be required to prevent rapid loss of surface moisture from the concrete. This can cause finishing difficulties and cracks in fresh concrete.

Spraying concrete with a curing compound.

You can prevent these problems by: (1) dampening the subgrade and forms; (2) minimizing the finishing time required by having sufficient manpower on hand; (3) erecting sunshades and windbreaks; (4) using temporary coverings, such as wet burlap or plastic sheeting during the finishing procedure, (5) using light fog sprays to prevent evaporation from the concrete; and (6) starting the curing process as soon as possible, using continuous wet methods or white pigmented curing compounds. During very hot, dry weather, place and finish concrete during the cooler, early morning or late evening hours.

Concrete Foundations and Floors

The function of a foundation is to provide a level and uniformly distributed support for a structure. In home construction, for example, the foundation must be strong enough to support and distribute the weight of the house, and it must be sufficiently level to prevent walls from cracking and doors and windows from sticking. The foundation also helps prevent cold air and dampness from entering the house. The two basic structural members of a foundation are footings and walls.

There are three basic types of foundations: the T-foundation, the slab foundation, and the pier or column foundation. The T-foundation consists of a trench footer, upon which a wall is placed. This type of foundation can be used above grade, on grade, or below grade. It is the strongest of all types, but it is the most expensive to build.

A slab foundation, as its name implies, is a solid slab of concrete that is poured directly on the ground. Footings can give extra support for the slab. Slab foundations require considerably less labor to build than most other foundation designs.

A pier or column foundation consists of individual footers, upon which the piers or columns are located. It is the simplest foundation but is generally used only for smaller structures.

All three types can be made of poured concrete, brick, concrete block, and stone. Columns can be steel or wood.

Footings

Footings act as the foundation base and transmit the superimposed load to the soil. The type and size of footings are usually determined by the soil condition and climate. In cold climates, footings must be below the frost line. Local codes usually establish this depth, which is often 4 feet or more.

Well-designed footings are important in preventing settling or cracks in the wall. For most soil conditions, the footing thickness or depth should be equal to the foundation wall thickness. Footings should project beyond each side of the wall one-half the wall thickness. The load capacity of the soil in some areas may require special designs. Local regulations specify these needs. This also applies to column and fireplace footings. If soil has a low load-bearing capacity, wider reinforced footings may be required.

The following rules apply to footing design and construction:

1. Footings must be a minimum of 6 inches thick.
2. If the footing excavation is too deep, fill it with packed, crushed stones or concrete—never replace the dirt.
3. Where soil is compact or claylike, and in areas with cold winter temperatures, a 4-inch base of sand and/or gravel is recommended.
4. Use form boards for footings where the soil condition prevents sharply cut trenches.
5. For load-bearing or freestanding walls higher than 4 feet, the footing should be below the frost line.
6. Reinforce footings with steel rods where they cross pipe trenches.

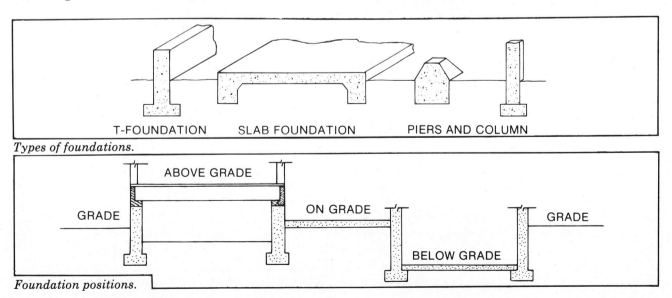

T-FOUNDATION   SLAB FOUNDATION   PIERS AND COLUMN

Types of foundations.

ABOVE GRADE

GRADE   ON GRADE   GRADE

BELOW GRADE

Foundation positions.

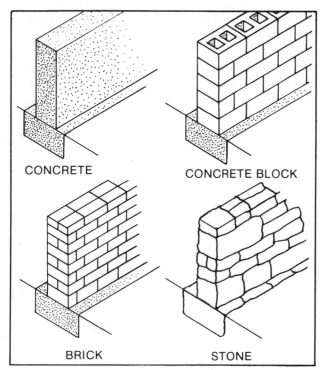

Foundation walls are constructed of concrete, concrete block, brick, and stone.

7. Use a key slot for better resistance to water entry at a wall location, especially in T-type foundations.
8. In freezing weather, cover with straw or supply heat to ensure proper curing.

Footings for piers, posts, or columns are generally square and include a pedestal on which the member will bear. A protruding steel pin is set in the pedestal to anchor a wood post. Bolts for the bottom plate of steel posts are usually set when the pedestal is poured. Steel posts are set directly on the footing, and the concrete floor is poured around them.

Pier or column footings vary in size depending on soil load capacity and the spacing of the piers, posts, or columns. Common sizes are 24 x 24 x 12 inches and 30 x 30 x 12 inches. The pedestal is sometimes poured after the footing. The minimum height should be about 3 inches above the finished basement floor and 12 inches above finish grade in crawl-space areas.

When using a slab foundation, footings need not be set so deeply into the ground. Check local building codes for information on how large and how deep they should be. The slab and its footing are generally poured in one mass; however, there has been a trend in recent years to pour the foundation and slab separately in colder climates. When doing this, an isolation or expansion joint is needed

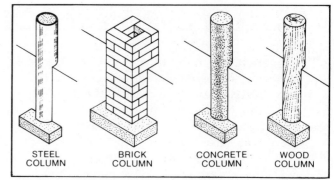

Columns are made of steel, brick, concrete, or wood.

to prevent excess stress between the foundation wall and the floor slab. This joint consists of a pre-molded strip of fiber material that extends the full depth of the slab or slightly below it. Some building codes require an integral footing-slab.

Regardless of foundation type, footings for fireplaces, furnaces, and chimneys are poured at the same time as other footings.

Laying Out the Footings and Foundation
Before starting construction, obtain a building permit from the proper officials, if required. Also check local codes for depth and thickness of footings and foundations, distance of building from property line, and other requirements.

The corners of a rectangular building must be perfectly square; so the foundation must be accurately laid out and carefully built. The dimensions of a building are measured along the outside of the foundation.

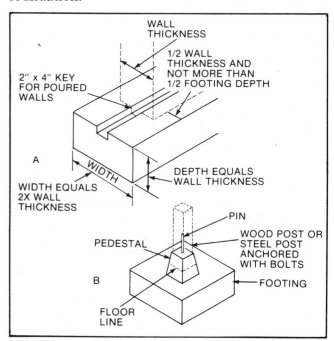

Concrete footings: (A) wall footing; (B) post footing.

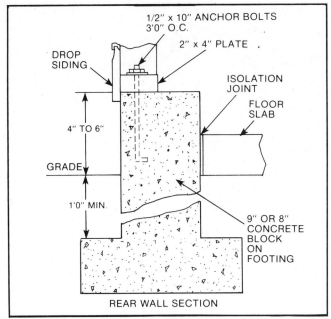

Slab and footing poured separately.

The first step in laying out a building is to indicate the location of the corners. If a surveying instrument is not available, you may proceed as follows:

1. Securely drive a heavy wooden stake where the highest corner of the building is to be located, as at A in the following illustration. Saw this stake off close to the ground. Drive nail X into the top of this stake to mark the exact location of the corner.

2. From nail X, stretch a string (dotted line AB) to stake M. The top of stake M should be higher than the top of stake A. String AB marks the line of one side of the foundation. This string should be 3 or 4 feet longer than the side of the

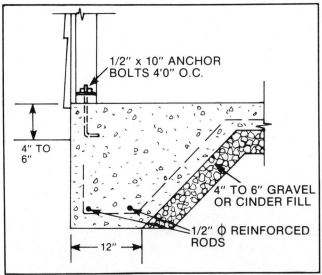

Integral footing-slab.

building. Fasten the string securely to M. All strings must be level.

3. Tie a second string (dotted line AD) to nail X. This string indicates one end of the building. Place this line so that it will be as nearly perpendicular to line AB as can be estimated. Place the temporary board K, which should be at least 4 to 6 feet long, so that it is crossed near its midpoint by line AD. The top edge of this board should be level with the top of stake A. Draw the string tight and temporarily fasten it.

4. From nail X, measure exactly 8 feet along line AB. Put a common pin through the string to mark the distance. Measure 6 feet from nail X along line AD. Put a pin through the string at this point. Prepare a narrow board or rigid rod with two points marked exactly 10 feet apart. Have one mark held directly under the intersection of AB and the pin (8 feet from X). Have the end of string AD swung to the right or left along the board K until the intersection AD at the pin (6 feet from X) is directly over the other, or the 10-foot mark on the rod. Lines AB and AD are now perpendicular. Secure line AD to board K.

5. Drive stakes B and D the proper distances from A to mark the length and width of the building. Saw these stakes off level with A and drive nails into them to mark the exact locations of the corners. Check the locations of these nails carefully as this determines the accuracy of the whole layout.

6. To locate the fourth corner of the rectangle, make distance BC equal to AD, and CD equal to AB. Mark C with a stake, then nail.

7. Check the rectangle for accuracy by measuring the diagonal distances. AC should equal BD. If these diagonal distances are not equal, check all measurements carefully and adjust the stakes and lines until the diagonals are equal, being sure to maintain the correct lengths of AB and AD.

8. Erect substantial batter boards, 5 to 8 feet from the corner stakes A, B, C, and D. The tops of the batter boards should be level with the top of the proposed foundation.

9. Using a plumb bob to insure accuracy, stretch string AB over line AB, and so on. The four points of intersection of the four strings will then be directly over the four nails that mark the location of the four corners. Mark the location of strings AB, BC, and others that may be needed on the batter boards, with nails or

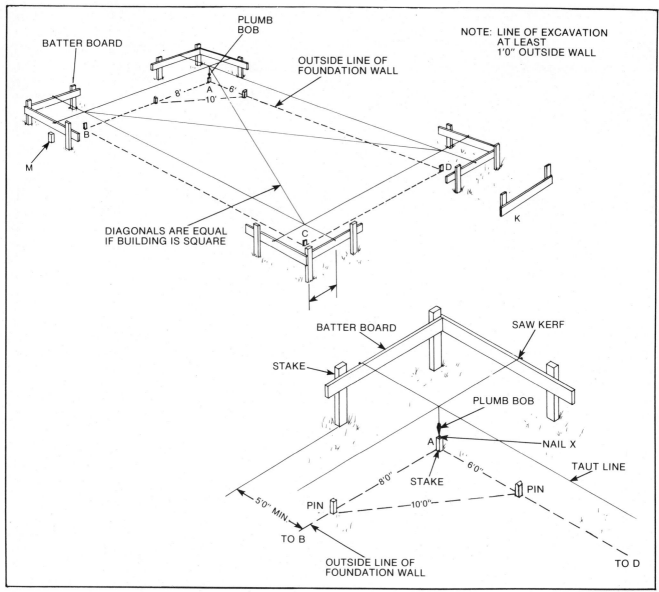

Method of staking and laying out a foundation.

saw cuts. The strings may then be easily and accurately replaced if dislodged.

Footing Forms If possible, the earth should form a mold for concrete wall footings. Otherwise, forms must be constructed. The sides of the forms are usually made of 2-inch lumber having a width equal to the depth of the footing. These boards are held in place with stakes and are maintained the correct distance apart by spreaders.

Conventional footings poured on a slope may slip downhill because of soil pressure. To prevent this sliding action, the footing should be constructed in a series of levels. Be sure each section is horizontal and below the frost line. The forms are constructed in the same manner as regular footings. Pour the horizontal sections of concrete first. After the concrete sets, pour the vertical sections.

Footings for columns and piers are usually square or rectangular. The four sides should be built and erected in panels, using form (duplex) nails to hold them together. Drive all nails from the outside, if possible, to make stripping easier. Anchoring bolts for wood or steel columns should be set when the footing is being poured.

The methods of mixing and pouring concrete are discussed fully in the first chapter. Before pouring the concrete, be sure the earth inside the forms is thoroughly moistened without making the soil really muddy.

The time required before removing the forms

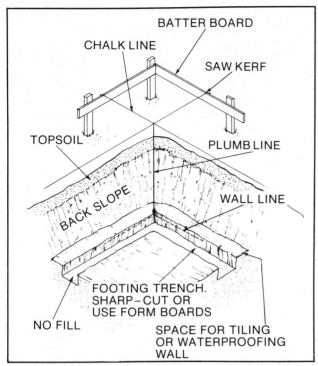

Establishing corners for an excavation and for digging the footing.

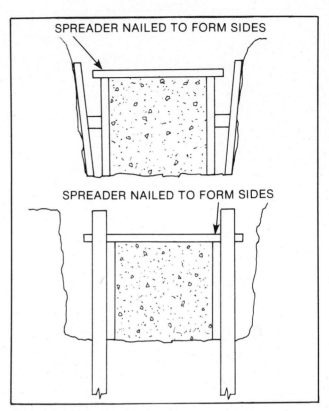

Methods of bracing footing forms.

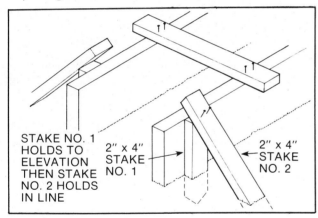

Typical wall footing forms.

depends on weather and humidity. In warm weather, it usually takes two days; in colder weather, it may take a week. When the forms are removed, the footing usually cures through the moisture in the surrounding soil.

Foundation Walls and Piers

Foundation wall heights and thicknesses are determined by the design and purpose of the structure. In home construction, foundation walls enclose the basement or crawl space and carry the various building loads. Wall thicknesses and types of construction are generally specified by local building regulations. Thicknesses of poured concrete basement walls may range from 6 to 10 inches, depending on story heights and length of unsupported walls.

Clear wall height should be no less than 7 feet from the finished basement floor surface to the bottom of the joists. Greater clearance will provide adequate headroom under girders, pipes, and ducts.

Poured Concrete Walls Poured concrete walls require tight, solid forming. Forming must be braced and tied to withstand the forces of the pouring operation and the fluid concrete. Poured concrete walls should be double-formed. For large jobs, reusable forms can usually be rented from contractor supply houses. Panels may consist of wood framing with plywood facings and are fasened together with clips or other ties. The formwork should be plumb, straight, and braced sufficiently.

Frames for cellar windows, doors, and other openings are set in place as the forming is erected along with forms for the beam pockets, which are located to support the ends of the floor beam. Reusable forms require sufficient blocking and bracing to keep them in place during pouring operations. Level marks of some type, such as nails along the form, should be used to assure a level foundation top to

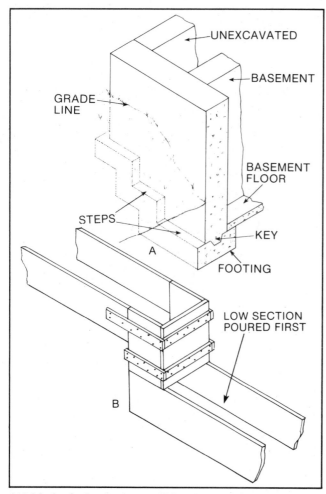

(A) Method of stepping wall footings in sloping ground.
(B) Typical form used for sloping ground footings.

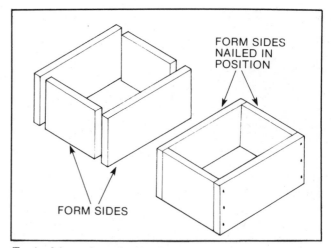

Typical form for column or pier footings.

provide a level sill plate for floor framing.

Before concrete is placed in forms which are to be stripped, the forms must be coated with a suitable oil or other lubricant to prevent bonding between the forms and the concrete. Almost any light-bodied petroleum oil is suitable except if finished concrete surfaces are to be painted.

For reusable forms a compound that prevents a bond and protects the form is necessary. On plywood forms, lacquer is satisfactory. If the forms are to be reused many times, painting helps to preserve them.

When the foundation is being laid, concrete should be poured without interruption and constantly puddled to remove any air pockets. The material should be worked under window frames and other blocking. Anchor bolts for the sill plate should be placed while the concrete is still plastic. Forms should not be removed until the concrete has hardened and acquired sufficient strength to support loads imposed during early construction. At least two days are required

when temperatures are well above freezing, and perhaps a week when outside temperatures are below freezing.

When building an addition to an old structure, secure a good joint where the new wall abuts against the old. If the walls are poured concrete, reinforcing rods can be set into the old walls. The new walls are poured around them to bind the two walls together. With concrete blocks, remove two or three units from the old wall and set in new ones so that they are half in the old and half in the new wall. This ties the two walls firmly together. Reinforcing rods can also be used here. The outside of the joint should be carefully sealed with plaster and tar to prevent water from seeping into the basement.

New poured concrete walls can be damp-proofed with one heavy coat of tar or asphalt. It should be applied to the outside surface from the footings to the finish grade line, but should not be applied until the surface of the concrete has hardened enough to assure good adhesion. In poorly drained soils, a membrane may be necessary.

Crawl-Space Foundations In many areas, crawl-space construction is preferred to a full basement or concrete slab, particularly for additions. The success of such a foundation depends on: (1) a good soil cover; (2) a small amount of ventilation; and (3) sufficient insulation to reduce heat loss.

A primary advantage of the crawl space over the full basement is reduced cost. Excavation or grading is required only for the footings and walls. In mild climates, the footings are located just below finish grade. In northern regions where frost penetrates deeply, the footing is often located 4 or more feet below finish grade. This requires more masonry work and increases the cost. Footings should always be poured over undisturbed soil and never over fill.

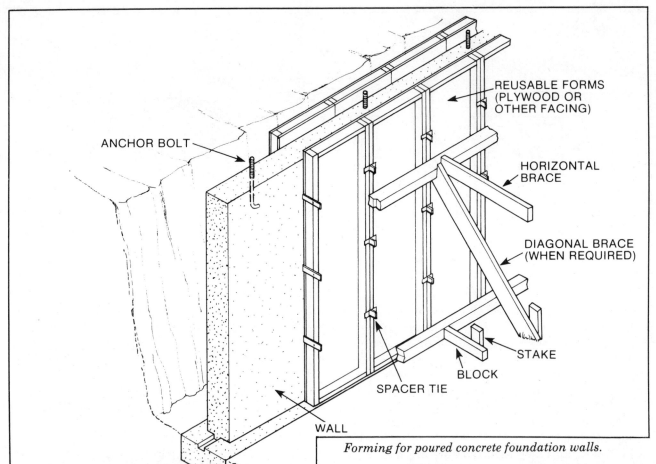

ANCHOR BOLT

REUSABLE FORMS
(PLYWOOD OR
OTHER FACING)

HORIZONTAL
BRACE

DIAGONAL BRACE
(WHEN REQUIRED)

STAKE

BLOCK

SPACER TIE

WALL

FOOTING

Forming for poured concrete foundation walls.

The construction of a masonry wall for a crawl space is much the same as for a full basement, except that no excavation is required within the walls. Waterproofing and drain tiles are usually not required. Footing size and wall thicknesses vary somewhat by location and soil conditions. A common minimum thickness for walls in single-story frame houses is 8 inches for hollow concrete block and 6 inches for poured concrete. The minimum footing thickness is 6 inches, and the width is 12 inches for concrete block and 10 inches for the poured foundation wall in crawl-space homes. In well-constructed houses, it is common practice to use 8-inch walls and 16 x 8-inch footings.

Poured concrete piers are often used to support floor beams in crawl-space houses. They should extend at least 12 inches above the groundline. Poured concrete piers should be at least 10 x 10 inches in size, with a 20 x 20 x 8 inch footing. Unreinforced concrete piers should be no greater in height than 10 times their smallest dimension. Spacing should not exceed 8 feet on center under exterior wall beams and interior girders set at right angles to the floor joists, and 12 feet on center under exterior wall beams set parallel to the floor joists. Exterior wall piers should not extend above grade more than four times their smallest dimension unless supported laterally by concrete walls. Pier footing sizes are based on the total load and the capacity of the soil.

Slab Foundations

A slab foundation is simply a concrete slab poured on top of a gravel bed with footings usually extending below the frost line. This type of foundation is common for basementless houses and is also used for home additions, garages, carports, sheds, and various other outbuildings.

Slab Foundations for Homes At one time the finish flooring for concrete floor slabs on the ground was asphalt tile adhered directly to the slab. These concrete floors did not prove satisfactory because the floors were cold and uncomfortable, and condensation occasionally collected on the floor. These problems were more common in colder climates.

Improvements in construction techniques have reduced the common problems of the slab floor,

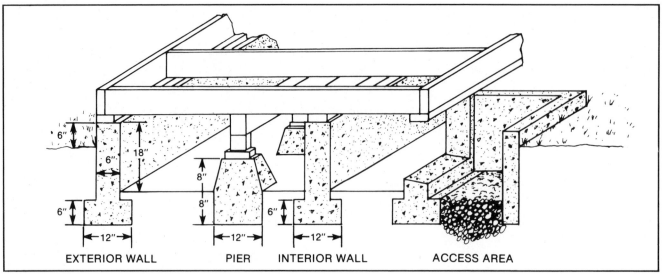

Typical crawl-space foundation construction.

but consequently increased the cost. Suitable insulation around the perimeter of the house helped to reduce the heat loss through the floor and around the walls. Radiant floor heating systems are effective in preventing cold floors and floor condensation problems. Peripheral warm-air heating ducts are also effective. Vapor barriers over a gravel fill under the floor slab prevent soil moisture from rising through the slab.

The following basic requirements are important in the construction of concrete floor slabs.

1. Establish the finish floor level high enough above the natural groundline so that the finish grade around the house can be sloped away for good drainage. The top of the slab should be at least 8 inches above the ground, and the siding at least 6 inches.
2. Topsoil should be removed, and sewer and water lines should be installed, then covered with 4 to 6 inches of gravel or crushed rock, well tamped in place.
3. A vapor barrier consisting of a heavy plastic film, such as 6-mil polyethylene, asphalt-laminated duplex sheet, or 45-pound or heavier roofing, with a minimum of ½-perm rating, should be used under the concrete slab. Joints should be lapped at least 4 inches and sealed. The barrier should be strong enough to resist puncturing during pouring.
4. A permanent, waterproof, nonabsorbent type of rigid insulation should be installed around the perimeter of the wall. Insulation may extend down on the inside of the wall vertically, or under the slab edge horizontally.
5. The slab should be reinforced with 6 by 6-inch No. 10 wire mesh. The concrete slab should be at least 4 inches thick.
6. After leveling and screeding, the surface should be floated with wood or metal floats while the concrete is still plastic. If a smooth, dense surface is needed for the installation of wood or resilient tile with adhesives, the surface should be steel troweled.

Concrete Slabs for Garages and Outbuildings Once the location of the garage or outbuilding has been determined, the next step is to lay out the corners of the slab. This can be accomplished by using the batter board method described earlier.

Remove enough earth so that the ground is 4 to 6 inches below the entire proposed floor level. Fill this floor area with 4 to 6 inches of gravel or cinders. Tamp it down or roll it firmly so that footsteps will not show. Level the gravel or cinders with a mason's level as a guide. To obtain a strong, durable floor, lay ⅜-inch tie rods or wire mesh in the concrete.

Forms should be constructed so that the foundation wall or top of the slab extends above the finished grade around the outside of the structure by 4 to 6 inches to protect the framing members from soil moisture. The slab perimeter should extend a minimum of 1 foot below grade. In colder regions, it may be necessary to extend the foundation wall to 3 or more feet below grade. The local building code may specify requirements.

The concrete slab and the footing slab may be poured simultaneously or separately. In either case, the finished surface of the slab should be leveled off with a strikeboard and then smoothed with a level float. For a gritty, nonskid surface, finish with a

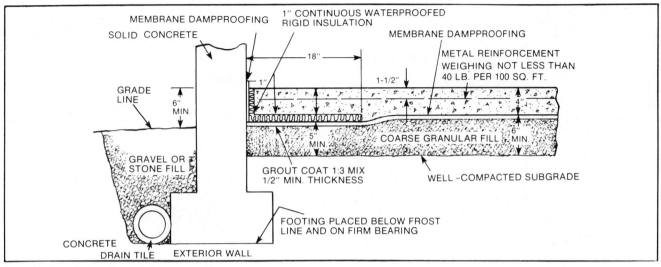

Details for slab-on-ground construction.

wood float. For a smooth surface, use a steel trowel. Finishing should only be done after the concrete has become quite stiff. After the finishing is completed, anchor bolts or clips should be set in the concrete to hold the wood sillplate. Be sure to anchor the sillplate in places where the studs or corners will not be nailed.

When tying a concrete slab to an existing structure, it is necessary to secure a good joint where the new wall abuts against the old.

Concrete Floors

Most concrete floors should be from 4 to 6 inches thick, depending on their use. For basements, a 4-inch thickness is usually adequate, but where heavy use is expected, the thickness should be 6 inches. Added strength can be provided by putting a heavy wire mesh or reinforcing rods in the concrete. Regardless of thickness, the slab should be uniform over its entire surface. Since some concrete floors are sloped for drainage, the subgrade under them must be sloped, too, for uniform floor thickness. Unless floors are poured in sandy subsoil, excavate 4 to 6 inches deeper and fill with cinders or gravel to prevent possible cracking or buckling because of moisture or freezing. This also helps waterproof the floor.

Where a pitch to the floor is desired to drain away laundry water, auto washing water, and so on, the slope should be at least ¼ inch to every foot. Before pouring a sloped floor, place the screeds or form side boards so that they slope toward the floor drains. In garages and barns, the slope should be toward the doorway. In either case, the line around the edge should be level.

After establishing the level line, place the screeds to slope away from it. The screeds should be held in place by nailing them to wood stakes driven in the subgrade. Once the screeds or guide boards are in place, the concrete can be poured around them. Strike off the concrete with a straightedge held level at one end and resting on the screed at the other. After the surface has been struck off, remove the screed from the wall edge and fill in with concrete.

When the concrete has begun to set, remove the guide boards, fill the channels left by the guides with fresh concrete, and trowel smooth. Joints are unnecessary in the floor itself.

As when pouring the foundation slab and footing separately, construct an expansion or isolation joint around the edges of the floors where they meet the side walls or foundation.

If trenches for sewers and drains are dug, they should be carefully backfilled and tamped to avoid future settlement. Settlement under a floor can cause cracks or breakage.

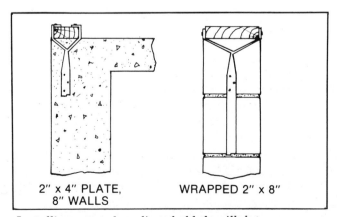

2" x 4" PLATE, 8" WALLS   WRAPPED 2" x 8"

Installing an anchor clip to hold the sillplate.

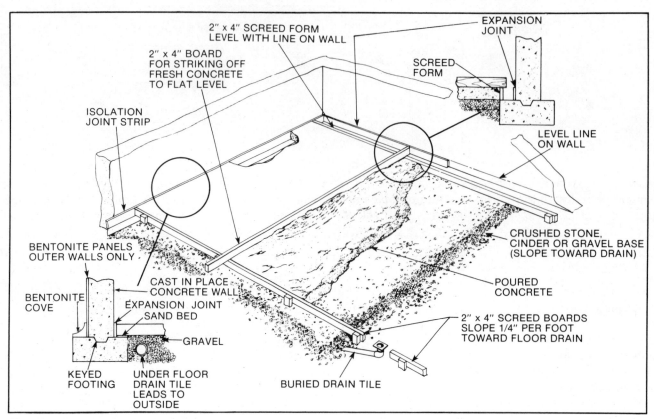

Details of pouring a concrete floor.

The surfaces of floors should be finished smooth for easy cleaning and proper water drainage. If wood or resilient flooring is to be laid directly on a concrete floor, the surface should be rough finished to increase the adhesion of the cement between the padding and the concrete floor.

Concrete Block Masonry

Concrete block combines the strength and durability of concrete with the ease of construction of masonry. While poured concrete remains the most popular material for floors and slabs, concrete block is often used in foundation walls in homes and above-ground walls for garages, sheds, barns, and industrial and commercial buildings. Concrete block is less expensive than other types of masonry, it is easier to build with than brick or stone, and it comes in hundreds of sizes, shapes, textures, and colors. Decorative concrete block has become increasingly popular for building fences, privacy screens, wall panels, and veneer.

Concrete Block: Sizes and Shapes

Concrete building units are made in sizes and shapes to fit different construction needs. Units are made in full and half-length sizes. Sizes are usually referred to by their nominal dimensions. A unit measuring 7⅝ inches wide, 7⅝ high, and 15⅝ inches long is referred to as an 8 x 8 x 16-inch unit. When it is laid in a wall with ⅜-inch mortar joints, the unit will measure exactly 16 inches long and 8 inches high.

Concrete masonry is much more than standard gray block. The concrete masonry fleur-de-lis design enhances this garden setting.

The corner unit is laid at a corner or where a smooth, rather than a recessed, end is required. The header unit is used in a backing course placed behind a brick face tier header course. Part of the block is cut away to admit the brick headers. The uses of the other special blocks shown in the illustration are self-evident.

The standard 8 x 8 x 16-inch concrete block weighs between 40 and 50 pounds. The normal-weight aggregates are sand and gravel, crushed stone, and air-cooled blast-furnace slag. Lightweight blocks, sometimes called cinder blocks, weigh about 25 to 30 pounds. The aggregates make the difference. Lightweight aggregates include expanded shale, clay, slate, coal, and cinders. Most codes require normal-weight blocks for foundation walls.

The texture of concrete block ranges from coarse to fine. Degrees of smoothness are achieved by changing the aggregate grading, mixing proportions, water content, and molding pressure. The distinctions of fine, medium, and coarse describe the relative smoothness of the surface. A fine texture is not only smooth but also has small, very closely spaced granular particles. A coarse texture has open pores because of the higher proportion of large-size aggregates in the mix.

If the concrete masonry surface will be a base for stucco or plaster, a coarse texture is desirable for the best bond. Coarse and medium textures provide sound absorption even when painted. Spray painting has proven to be the most effective method of application. A fine texture, while not as sound absorbent, is easier and more economical to paint.

Modular Planning Concrete masonry walls should be carefully planned to make maximum use of full- and half-length units, thus minimizing cutting and fitting on the job. Length and height of the wall, width and height of openings and wall areas between doors, windows, and corners should be planned to use standard full-size and half-size units. All horizontal dimensions should be in multiples of nominal full-length masonry units, and both horizontal and vertical dimensions should be designed to be in multiples of 8 inches. The following tables list the nominal lengths of concrete masonry walls by stretchers and the nominal heights of concrete masonry walls by courses.

When units 8 x 4 x 16 are used, the horizontal dimensions should be planned in multiples of 8 inches (half-length units) and the vertical dimensions should be planned in multiples of 4 inches. If the thickness of the wall is greater or less than the length of a half unit, a special length unit is required at each corner in each course.

For nearly any masonry project, concrete block or brick, a footing is required to prevent shifting caused by normal ground movement. The footing

Nominal Length of Concrete Masonry Walls by Stretchers		
	Nominal Length of Concrete Masonry Walls	
No. of Stretchers	**Units 15⅝″ long and half units 7⅝″ long with ⅜″ thick head joints.**	**Units 11⅝″ long and half units 5⅝″ long with ⅜″ thick head joints.**
1	1′4″	1′0″
1½	2′0″	1′6″
2	2′8″	2′0″
2½	3′4″	2′6″
3	4′0″	3′0″
3½	4′8″	3′6″
4	5′4″	4′0″
4½	6′0″	4′6″
5	6′8″	5′0″
5½	7′4″	5′6″
6	8′0″	6′0″
6½	8′8″	6′6″
7	9′4″	7′0″
7½	10′0″	7′6″
8	10′8″	8′0″
8½	11′4″	8′6″
9	12′0″	9′0″
9½	12′8″	9′6″
10	13′4″	10′0″
10½	14′0″	10′6″
11	14′8″	11′0″
11½	15′4″	11′6″
12	16′0″	12′0″
12½	16′8″	12′6″
13	17′4″	13′0″
13½	18′0″	13′6″
14	18′8″	14′0″
14½	19′4″	14′6″
15	20′0″	15′0″
20	26′8″	20′0″

(Actual length of wall is measured from outside edge to outside edge of units and is equal to the nominal length minus ⅜″ [one mortar joint].)

Nominal Height of Concrete Masonry Walls by Courses		
	Nominal Height of Concrete Masonry Walls	
No. of Courses	**Units 7⅝″ high and ⅜″ thick bed joint.**	**Units 3⅝″ high and ⅜″ thick bed joint.**
1	8″	4″
2	1′4″	8″
3	2′0″	1′0″
4	2′8″	1′4″
5	3′4″	1′8″
6	4′0″	2′0″
7	4′8″	2′4″
8	5′4″	2′8″
9	6′0″	3′0″
10	6′8″	3′4″
15	10′0″	5′0″
20	13′4″	6′8″
25	16′8″	8′4″
30	20′0″	10′0″
35	23′4″	11′8″
40	26′8″	13′4″
45	30′0″	15′0″
50	33′4″	16′8″

(For concrete masonry units 7⅝″ and 3⅝″ in height laid with ⅜″ mortar joints. Height is measured from center to center of mortar joints.)

should be cast of quality concrete on firm ground below the frost line. Unless local codes stipulate otherwise, it is general practice to make footings for small buildings twice the width of the wall thickness. The thickness of the footings is equal to one-half their width.

Masonry Tools and Equipment Before you start any masonry project, be sure you have the necessary tools close at hand. For projects using concrete blocks, the mason's tools include a trowel, bolster, hammer, and jointer.

The mason's trowel may be a brick, a buttering, or a pointing trowel. The trowel is used for mixing, placing, and spreading mortar. The hammer is used for tapping masonry units into the beds where necessary and for chipping and rough-cutting. For smoother cutting, the bolster (also called a brick-cutting chisel or brick set), is used. Breaking into bats and closures is done with the chisel peen on the mason's hammer. Splitting and rough-breaking is done with the head or flat of the hammer.

The jointer, of which there are several types, is

used for making various joint finishes which will be described later.

Maintain a constant check on your courses to ensure that they are level and plumb. The equipment for this vital purpose consists of a length of line, a steel square and level, and a straightedge. The square is used to lay out corners and for other right-angle work. The mason's level is used like the carpenter's level in wood construction. The straightedge is used in conjunction with the level for leveling or plumbing long stretches.

A mortarboard for holding a supply of ready-to-use mortar should be constructed as shown. If mortar is to be mixed by hand, it should be mixed in a mortar box. The box can be lined with metal to prolong the life of the box. Other required equipment includes shovels, mortar hoes, wheelbarrows, and buckets.

Mortar

Good mortar is necessary for quality workmanship and proper structural performance of concrete

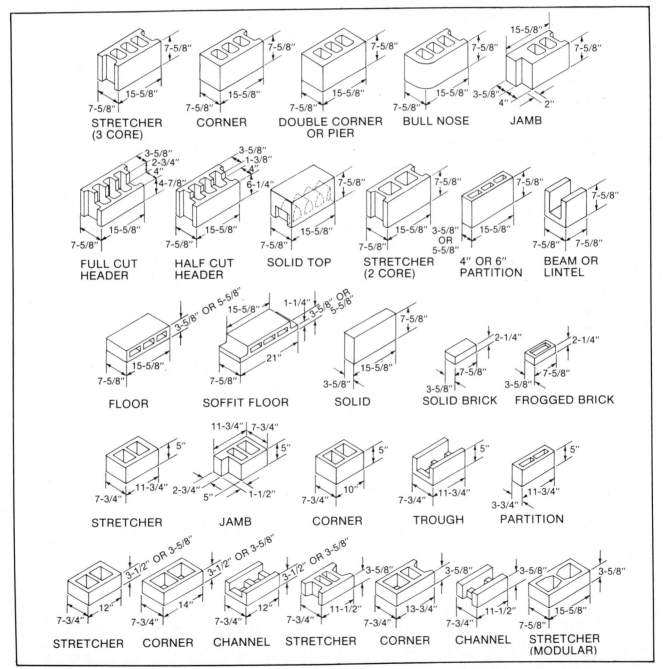

Typical sizes and shapes of concrete masonry block units.

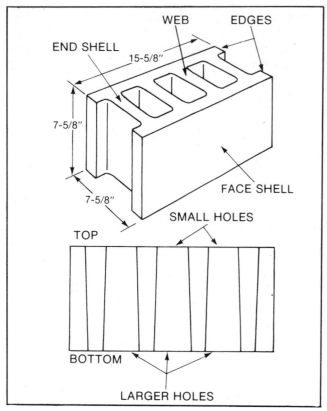

Concrete block terminology.

masonry. Since mortar bonds masonry units into strong, durable, weathertight structures, it has many desirable properties. Materials must comply with specifications. Desirable properties include workability, water retentivity, and a consistent rate of hardening.

There are several different mortar mixes. You can substitute regular portland cement with masonry cement, which contains the proper percentage of

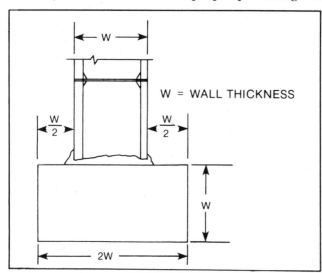

Dimensions of masonry wall footings.

lime. Masonry cement reduces preparation time and is more convenient to use, but it is more expensive. Both cements produce mortar with acceptable properties.

The following table is a guide to the selection of mortar types. For most projects in this book, type M or S can be used.

Guide to the Selection of Mortar Type	
Kind of Masonry	**Types of Mortar***
Foundations:	
Footings	M, S
Walls of solid units	M, S, N
Walls of hollow units	M, S
Hollow walls	M, S
Masonry other than foundation masonry:	
Piers of solid masonry	M, S, N
Piers of hollow units	M, S,
Walls of solid masonry	M, S, N, O
Walls of solid masonry, other than parapet walls, not less than 12 inches thick or more than 35 feet in height, supported laterally at intervals not exceeding 12 times the wall thickness	M, S, N, O, K
Walls of hollow units, load-bearing or exterior, and hollow walls 12 inches or more in thickness	M, S, N
Hollow walls, less than 12 inches thick where assumed design wind pressure:	
1. exceeds 20 psf	M, S,
2. does not exceed 20 psf	M, S, N
Linings of existing masonry, either above or below grade	M, S,
Masonry other than above	M, S, N

* *More on types of mortar can be found on page 38.*

Architectural effects that add color contrast or harmony to the joints between units can be obtained by using white or pigmented mortars. White mortar is made with white masonry cement, or with white portland cement and lime, and white sand. Use white masonry cement or white portland cement with colored mortars to produce cleaner, brighter colors and for making pastel colors such as buff, cream, ivory, pink, and rose. Integrally colored mortar may be obtained by using color pigments, colored masonry cements, or colored sand. Brilliant or intense colors are not generally attainable in masonry mortars. The color of the mortar joints also depends on the cementitious materials, aggregate, and water-cement ratio.

Mixing Mortar Sufficient mixing water should be added to obtain the desired consistency. If a large quantity of mortar is required, it should be mixed in a drum-type mixer similar to those used for mixing concrete. Mixing time should not be less than three minutes. All dry ingredients should be placed in the mixer first and mixed for at least one minute before adding the water.

Unless large amounts of mortar are required, mortar is mixed by hand using a mortar box. Mix all the ingredients thoroughly to obtain a uniform texture. All dry material should be mixed first, working from one end of the box to the other with a hoe. Add two-thirds to three-fourths of the required water and mix with the hoe until the batch is uniformly wet. Additional water is added carefully while mixing until the desired workability is attained. Allow the batch to stand for approximately five minutes and then thoroughly remix it with the hoe. A steel drum filled with water should be kept close to the mortar box for the water supply. A second drum of water should be available for shovels and hoes when not in use.

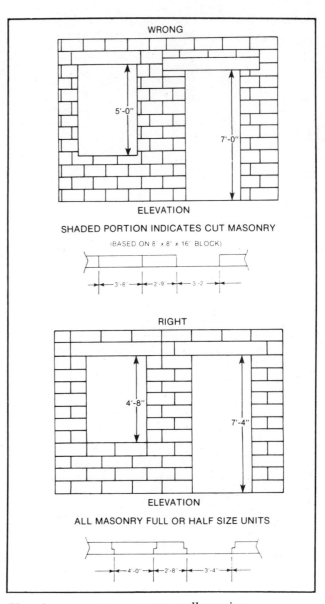

Planning a concrete masonry wall opening.

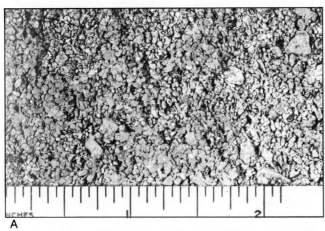

(A) Medium and (B) coarse textures in concrete block.

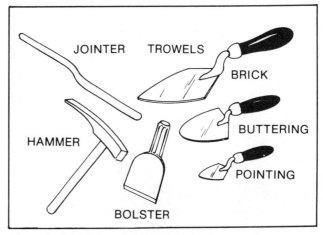

Masonry tools.

Proportion Specifications for Mortar				
		Parts by volume		
Specification	Mortar type	Portland cement or portland blast-furnace slag cement	Masonry cement	Hydrated lime or lime putty
For plain masonry	M	1	1	—
		1	—	¼
	S	½		—
		1	—	Over ¼ to ½
	N	—	1	—
		1	—	Over ½ to 1¼
	O	—	1	—
		1	—	Over 1¼ to 2½
	K	1	—	Over 2½ to 4
For reinforced masonry	PM	1	1	—
	PL	1	—	¼ to ½

Fresh mortar should be prepared at the rate it is used for uniform workability. Mortar that has been mixed but not used immediately dries out and stiffens. Loss of water by absorption and evaporation on a dry day can be reduced by wetting the mortarboard and covering the mixed mortar.

To restore workability, mortar may be retempered by adding water and remixing thoroughly. Additional water may slightly reduce the compressive strength of the mortar, but the end effect is acceptable. Plastic mortar has a better bond strength than dry, stiff mortar.

Mortar that has stiffened because of hydration should be discarded. Since it is difficult to determine by sight or feel whether mortar stiffening is due to evaporation or hydration, mortar should be used within 2½ hours after mixing.

If colored mortar is used, retempering may cause a significant lightening of the mortar.

Estimating the Required Materials   Use rule 38, as described below, for calculating the amount of raw material needed to mix 1 yard of mortar.

Builders have found that it requires about 38 cubic feet of raw materials to make 1 cubic yard of mortar. In using the 38 rule, take the rule number and divide it by the sum of the quantity figures specified in the mix. For example, building specifications call for a 1:3 mix for mortar and fine aggregates: $1 + 3 = 4$. Then, $38 \div 4 = 9½$. You will need 9½ sacks or 9½ cubic feet of cement. In order to calculate the amount of fine aggregates (sand), you simply multiply 9½ by 3. The product, 28½ cubic feet, is the amount of sand you need to mix 1 cubic yard of mortar using a 1:3 mix. The sum of the two required quantities should always equal 38.

Once you have ordered materials, think about storing the block. Since most projects take more than one day to complete, you will have to keep the stored blocks dry. Stack blocks on planks or platforms set above the ground, located as close as possible to the construction site. Cover with polyethylene sheets, roofing paper, or another waterproof cover. The unfinished top of a concrete masonry structure should be covered with tarpaulins or boards to prevent rain or snow from falling inside the unit cores.

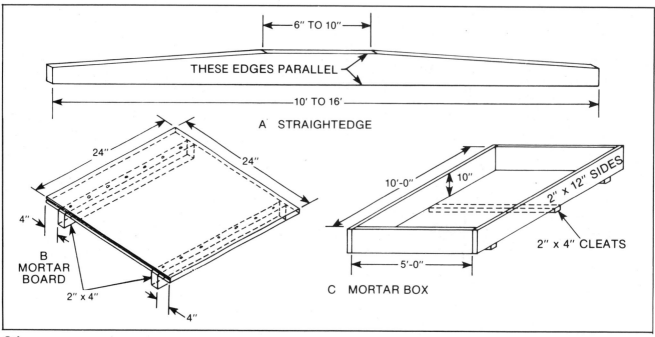

Other masonry equipment.

1. *To check the layout, you should string out the block without mortar.*

2. *A full mortar bed is necessary for laying the first course.*

3. *Positioning and leveling the corner block.*

4. *Blocks are buttered for vertical joints.*

Laying Concrete Block

To begin laying concrete block, locate the corners of the wall and check the layout by stringing out the blocks for the first course without mortar. Lay out the block, leaving space for the mortar. A chalk snapline is used to mark the footing and align the block accurately. A full bed of mortar is spread and furrowed with the trowel to ensure plenty of mortar along the bottom edges of the face shells of the block for the first course. The corner block should be laid first and carefully positioned. Lay all block with the thicker end of the face shell up, to provide a larger mortar-bedding area. Mortar is applied only to the ends of the face shells for vertical joints. Place several blocks on end and apply the mortar to the vertical face shells at one time. Each block is brought to the site and pushed downward into the mortar bed and against the previously laid block to obtain a well-filled vertical mortar joint. After three or four blocks have been laid, the mason's level is used as a straightedge to assure correct alignment of the blocks. Then, the blocks are carefully checked with the level and brought to proper grade and made plumb by tapping with the trowel handle. The first course of concrete masonry should be laid with great care, to assure that succeeding courses, and finally the wall, are straight and true.

After the first course is laid, mortar is applied only to the horizontal face shells of the block. Mortar for the vertical joints may be applied to the vertical face shells of the block to be placed or to the block previously laid. The corners of the wall are built first, usually four or five courses higher than the center of the wall. As each course is laid at the corner, it is checked with a level for alignment, for levelness, and plumbness. Each block is carefully checked with a level or straightedge to make certain that the block faces are all in the same plane. The use of a story or course pole, a board with markings 8 inches apart, provides an accurate method of determining the top of the masonry for each course. Joints are ⅜ inch thick. In building the corners, each course is stepped back a half block. You should check the horizontal spacing of the block by placing your level diagonally across the corners of the block.

When filling in the wall between the corners, a mason's line is stretched from corner to corner for each course, and the top outside edge of each block is laid to this line.

Gripping the block is important. It should be tipped slightly towards you so you can see the edge of the course below. Place the lower edge of the block directly over the course below. All adjustments to final position must be made while the mortar is soft and plastic. Adjustments made after the mortar has stiffened will break the mortar bond and allow the penetration of water. Each block is leveled and aligned to the mason's line by tapping lightly with the trowel handle. Each course is checked with the mason's level to ensure straight walls.

To assure a good bond, do not spread mortar too

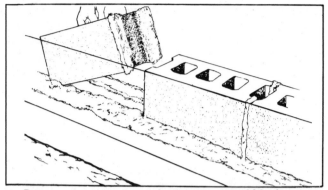

5. *Proper way to position the first course of blocks.*

6. *Leveling the block.*

7. *Plumbing the block.*

8. *Making the vertical joints.*

9. *Aligning each course.*

10. *Leveling each course at the corner.*

11. *Plumbing the courses at the corner.*

far ahead of actually laying the block. As each block is laid, excess mortar at the joints is cut off with the trowel and is thrown back on the mortarboard to be reworked into the fresh mortar.

When installing the closure block, all edges of the opening and all four vertical edges of the closure block are buttered with mortar. If any of the mortar falls out leaving an open joint, the block should be removed and the procedure repeated.

Weathertight joints and neat appearance require proper tooling. Mortar joints should be tooled after a section of the wall has been laid and the mortar has become thumbprint hard. Tooling compacts the mortar and forces it tightly against the masonry unit. Horizontal joints should be tooled first, followed by striking the vertical joints with a small S-shaped jointer. Any remaining mortar burrs should be trimmed flush with the face of the wall with a trowel or removed by rubbing with a burlap bag or soft-bristle brush.

Wood plates are fastened to the top course by anchor bolts ½ inch in diameter, 18 inches long, and spaced not more than 4 feet apart. The bolts are placed in the cores of the top two courses of block. Cores are then filled with concrete or mortar. Pieces of metal lath, placed under the cores to be filled, will hold the concrete or mortar filling in place. The threaded end of the bolt should extend a few inches above the top of the wall.

Cutting Concrete Block It is sometimes necessary to cut a block to fit a particular location. This

12. *Using a story pole to check the spacing of courses.*

13. *Diagonal check of the horizontal spacing of units.*

14. *A mason's line is used to lay blocks between corners.*

15. *Adjusting the block between corners.*

16. *It is necessary to cut off excess mortar.*

17. *Installing the closure block.*

18. *Tooling mortar joints. Horizontal (top) and vertical (bottom).*

can be done by scoring the block with a bolster and then breaking it along the score lines.

Patching and Cleaning Block Walls Patching mortar joints or filling holes should be done with fresh mortar.

Hardened, embedded mortar smears cannot be removed and paint may not hide smears, so prevent smearing mortar onto the surface of the block. Concrete block walls should not be cleaned with an acid wash. Mortar droppings should be allowed to dry before removal with a trowel. Most of the mortar can be removed by rubbing with a small piece of broken block after the mortar is dry and hard. Brushing the rubbed spots will remove most of the mortar.

Mortarless Concrete Block Walls Recently it has become possible to construct concrete block walls without mortar due to the introduction of a portland cement-based blend of fiberglass and a fine sand mixture. This mixture permits easy troweling, leaving an attractive, waterproof, stuccolike finish that can be treated for decorative effects.

When laying the block using this bonding system, set only the first course of blocks in a bed of mortar as described earlier. After the first course is in place, begin dry-stacking blocks in a staggered pattern. Remove all burrs and chips from the blocks to ensure a close fit. Recheck the level to ensure even height all the way around every few courses. Shim, if necessary, using sand or dry mortar mix. Use corner blocks at all corners. Stack all of the blocks before troweling on the cement-fiberglass mixture.

19. Tooled mortar joints for weathertight construction.

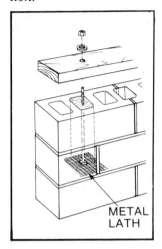

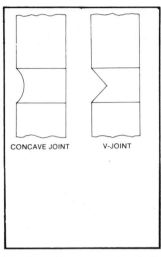

CONCAVE JOINT   V-JOINT

METAL LATH

20. Installing an anchor bolt on the top of the wall.

Dimensions of Dry-Stacked Blocks		
(Standard block units 7⅝″ × 7⅝″ × 15⅝″)		
Number of Blocks	Length (Laid End to End)	Height* (Stacked)
1	1 ft., 3⅝ in.	0 ft., 8 in.
2	2 ft., 7¼ in.	1 ft., 3⅝ in.
3	3 ft., 10⅞ in.	1 ft., 11¼ in.
4	5 ft., 2½ in.	2 ft., 6⅞ in.
5	6 ft., 6⅛ in.	3 ft., 2½ in.
6	7 ft., 9¾ in.	3 ft., 10⅛ in.
7	9 ft., 1⅜ in.	4 ft., 5¾ in.
8	10 ft., 5 in.	5 ft., 1⅜ in.
9	11 ft., 8⅝ in.	5 ft., 9 in.
10	13 ft., ¼ in.	6 ft., 4⅝ in.
12	15 ft., 7½ in.	7 ft., 7⅞ in.
15	19 ft., 6⅜ in.	9 ft., 6¾ in.

*Approximately 56 blocks are required per 50 square feet of wall. One 50-lb. bag of cement fiberglass mix covers 50 square feet with a ⅛-inch coating. *Includes ⅜-inch mortar bed, first course only.*

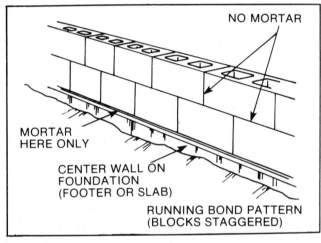

NO MORTAR

MORTAR HERE ONLY

CENTER WALL ON FOUNDATION (FOOTER OR SLAB)

RUNNING BOND PATTERN (BLOCKS STAGGERED)

Details of mortarless concrete block wall.

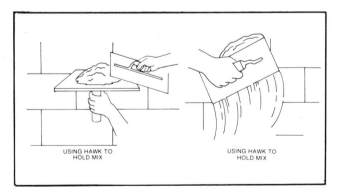

USING HAWK TO HOLD MIX

USING HAWK TO HOLD MIX

How the finish is given to a mortarless block structural wall.

The dimensions of dry-stacked blocks are given in the following table.

Before applying the cement-fiberglass mix, uniformly wet the concrete block wall with water to prevent excessive absorption of water from the mix. If the surface dries before application, rewet it. Avoid saturating blocks. It is important that all of the blocks be completely free of paint, oil, dirt, dust, or foreign materials which would interfere with bonding.

Mix the material according to the directions on the container. Starting at the bottom, and using a finishing trowel in an upward sweeping motion, completely cover the wall surface with the mixture to a minimum thickness of ⅛ inch. Apply the mixture to both sides of the wall. If there is to be an interruption of more than one hour during application, work should cease in a straight edge at a point at least 2 inches from any joint.

After 8 hours, but prior to 24 hours following application of the cement-fiberglass compound, dampen the wall with a water mist. Spray it once or twice a day for several days. Additional spraying may be required in hot, dry, or windy weather. Do not apply the mist at temperatures below 40 degrees Fahrenheit, and protect it from freezing for 48 hours. Like any other cement mix, this new material has greater strength when moist cured. It provides maximum strength in about 28 days. Check local codes for requirements and restrictions before undertaking any size mortarless construction.

Brick and Stone Masonry

Brick masonry units are of uniform size, are small enough to be placed with one hand, and are laid in courses with mortar joints to form walls of virtually unlimited length and height. Bricks are kiln-baked from various clay and shale mixtures. The chemical and physical characteristics of the ingredients vary considerably. The ingredients and kiln temperatures combine to produce brick in a variety of colors and hardnesses.

The dimensions of a United States standard building brick are 2½ x 3¾ x 8 inches. The actual dimensions of brick may vary a little because of shrinkage during baking.

Frequently, you must cut a brick into various shapes. They are called half or bat, three-quarter closure, quarter closure, king closure, queen closure, and split. They are used to fill in the spaces at corners and such other places where a full brick will not fit.

The six surfaces of a brick are called the face, the side, the cull, the end, and the beds.

A finished brick structure contains face brick placed on the exposed face of the structure and backup brick placed behind the face brick. Face brick is often of higher quality than the backup brick. The entire wall may be built of common brick which is made from pit-run clay, with no attempt at color control and no special surface treatment like glazing or enameling. Most common brick is red.

Although any surface brick is a face brick as distinguished from a backup brick, the term face brick is also used to distinguish high-quality brick from brick which is of common brick quality or less. Face brick is more uniform in color than common brick, and it may be available in a variety of colors. It may have a special surface finish and has a better surface appearance than common brick. It may also be more durable because of materials, or as a result of special manufacturing methods.

Backup brick may consist of brick which is inferior in quality even to common brick. Brick which has been underburned or overburned or brick made with inferior clay or by inferior methods is often used for backup brick.

Another type of classification divides brick into grades in accordance with the probable climatic conditions to which it will be exposed.

- Grade SW is brick designed to withstand exposure to the below-freezing temperatures in a moist climate common to the northern regions of the United States.
- Grade MW is brick designed to withstand exposure to the below-freezing temperatures in a drier climate.
- Grade NW is brick primarily intended for interior or backup brick. It may be exposed in regions where no frost action occurs or in regions where frost action occurs but the annual rainfall is less than 15 inches.

Types of Brick

There are several types of brick. Some are different in formation and composition, while others vary according to use. Commonly used types of brick are:

Building brick Formerly called common brick, it is made of ordinary clays or shales and burned in kilns. These bricks do not have special scorings or markings and are not produced in any special color or surface texture. Building brick is also known as hard and kiln run brick. It is generally used for the backing courses in solid or cavity brick walls.

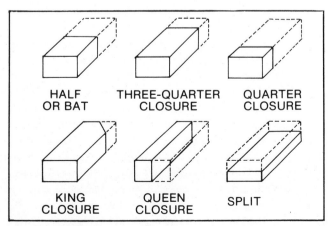

Common shapes of cut brick.

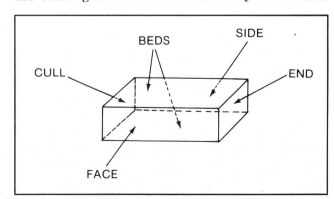

Names of brick surfaces.

The harder and more durable kinds are preferred.

Face brick This is used in the exposed face of a wall and is of higher quality than building brick. It has better durability and appearance. Common colors of face brick are shades of brown, red, gray, yellow, and white.

Clinker bricks This type of brick is overburned in the kiln and is usually hard and durable and may be irregular in shape. Rough hard corresponds to the clinker classification.

Press brick The dry press process creates this brick which has regular smooth faces, sharp edges, and perfectly square corners. Generally press brick is used as face brick.

Glazed brick This brick has one surface of each brick glazed in white or another color. The ceramic glazing consists of mineral ingredients which fuse together in a glasslike coating during burning. This type of brick is particularly suited for walls or partitions in hospitals, dairies, laboratories, or other buildings where cleanliness and ease of cleaning is essential.

Fire brick This type is made of a special fire clay which will withstand the high temperatures of fireplaces, boilers, and similar usages without cracking or decomposing. Fire brick is generally larger than regular structural brick and often it is hand-molded.

Cored bricks These are bricks made with two rows of five holes extending through their beds to reduce weight. There is no significant difference between the strength of walls constructed with cored brick and those constructed with solid brick. Resistance to moisture penetration is about equal.

Sand-lime bricks These are made from a lean mixture of slaked lime and fine silicious sand molded under mechanical pressure and hardened under steam pressure.

Mortar for Brick Masonry

Mortar is used to bond the brick together and, unless properly mixed and applied, will be the weakest part of brick masonry. The strength and resistance to rain penetration are dependent on the strength of the bond. Water is essential to the development of bond, and if the mortar contains insufficient water, the bond will be weak and spotty. Irregularities in brick dimensions and shape are corrected by the mortar joint.

Mortar should be plastic enough to work with a trowel. The properties of mortar depend largely upon the type of sand used in it. Clean, sharp sand produces excellent mortar. Too much sand causes

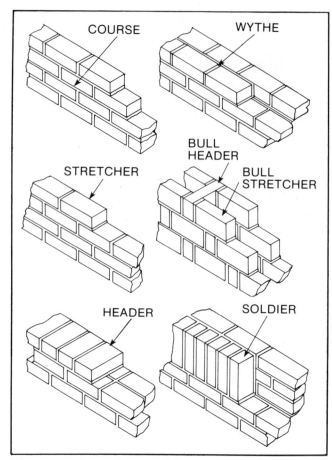

Masonry units and mortar joints.

mortar to segregate, drop off the trowel, and weather poorly.

The selection of mortar for brick construction depends on the use of the structure. For example, the recommended mortar for use in laying up interior nonload-bearing partitions would not be satisfactory for foundation walls.

- Mortar Type M is suitable for general use and is recommended specifically for masonry below grade and in contact with earth, such as foundations, retaining walls, and walks.
- Mortar Type S is suitable for general use and is recommended where high resistance to lateral forces is required.
- Mortar Type N is suitable for general use in exposed masonry above grade and is recommended specifically for exterior walls subjected to severe exposures.
- Mortar Type O is recommended for load-bearing walls of solid units where the compressive stresses do not exceed 100 pounds per square inch and the masonry will not be subjected to freezing and thawing and the presence of excessive moisture.

Ordering Brick and Mortar

Planning brick areas is similar to the procedure described for concrete block walls. If you have flexibility in planning the length and the height of the wall, assign multiples of nominal brick dimensions to avoid having to cut or use half-height bricks. The resulting wall will look more uniform and pleasing to the eye.

Actual dimensions of frequently used brick are:

Common	2¼ × 3¾ × 8
Modular	2¼ × 3⅝ × 7⅝
Jumbo	2¾ × 3¾ × 8
Norman	2¼ × 3⅝ × 11⅝
SCR	2⅛ × 5½ × 11½
Roman	1⅝ × 3⅝ × 11⅝
Baby Roman	1⅝ × 3⅝ × 7⅝
Fire Brick	2½ × 3⅝ × 9
Oversize	Sizes vary with manufacturer

Nominal, or working, dimensions of the brick equal the actual dimension plus the width of the mortar joint. For example, the nominal dimension for the common 2¼ x 3¾ x 8-inch brick, using ½-inch mortar joints, would be 2¾ x 4¼ x 8½ inches. While common brick is usually laid up with ½-inch joints, ⅜- or ⅝-inch joints can be used. Certain bricks such as the modular or Roman bricks are specifically designed to be laid up with ⅜-inch joints. Nominal dimensions change accordingly.

Divide the proposed length and height of the structure by the nominal dimension of the brick. Then, vary the figures until you can divide the numbers by whole bricks vertically and by whole and half-bricks horizontally. To make calculations easier, the following table lists the wall heights possible using 2¼-inch high brick with ½-inch joints. To use the table, find the nearest height to the desired one, then look at the corresponding number of courses listed next to it. For example, if you want to build a wall about 8 feet high with 2¼-inch bricks and ½-inch mortar joints, 35 courses of brick would bring the wall to 8 feet, ¼-inch.

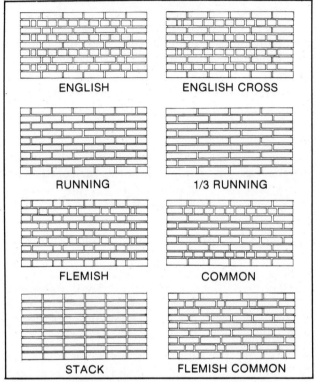

Some types of brick masonry bond.

	Height of Courses—2¼-Inch Brick, ½-Inch Joint										
Courses	Height	Courses	Height	Courses	Height	Courses	Height	Courses	Height		
1	0'2¾″	21	4'9¾″	41	9'4¾″	61	13'11¾″	81	18'6¾″		
2	0'5½″	22	5'0½″	42	9'7½″	62	14'2½″	82	18'9½″		
3	0'8¼″	23	5'3¼″	43	9'10¼″	63	14'5¼″	83	19'0¼″		
4	0'11″	24	5'6″	44	10'1″	64	14'8″	84	19'3″		
5	1'1¾″	25	5'8¾″	45	10'3¾″	65	14'10¾″	85	19'5¾″		
6	1'4½″	26	5'11½″	46	10'6½″	66	15'1½″	86	19'8½″		
7	1'7¼″	27	6'2¼″	47	10'9¼″	67	15'4¼″	87	19'11¼″		
8	1'10″	28	6'5″	48	11'0″	68	15'7″	88	20'2″		
9	2'0¾″	29	6'7¾″	49	11'2¾″	69	15'9¾″	89	20'4¾″		
10	2'3½″	30	6'10½″	50	11'5½″	70	16'0½″	90	20'7½″		
11	2'6¼″	31	7'1¼″	51	11'8¼″	71	16'3¼″	91	20'10¼″		
12	2'9″	32	7'4″	52	11'11″	72	16'6″	92	21'1″		
13	2'11¾″	33	7'6¾″	53	12'1¾″	73	16'8¾″	93	21'3¾″		
14	3'2½″	34	7'9½″	54	12'4½″	74	16'11½″	94	21'6½″		
15	3'5¼″	35	8'0¼″	55	12'7¼″	75	17'2¼″	95	21'9¼″		
16	3'8″	36	8'3″	56	12'10″	76	17'5″	96	22'0″		
17	3'10¾″	37	8'5¾″	57	13'0¾″	77	17'7¾″	97	22'2¾″		
18	4'1½″	38	8'8½″	58	13'3½″	78	17'10½″	98	22'5½″		
19	4'4¼″	39	8'11¼″	59	13'6¼″	79	18'1¼″	99	22'8¼″		
20	4'7″	40	9'2″	60	13'9″	80	18'4″	100	22'11″		

To determine how many bricks and how much mortar is needed to complete a brick area, use the following table. For example, the approximate dimensions of a wall are 8 feet by 10 feet. The area of the wall is equal to 80 square feet. If the wall is a single brick thick, the required number of bricks and mortar would be: 8 times 61.7, or 494 bricks, and 8 times 0.8, or about 6½ cubic feet of mortar. Add an additional 10 percent to the total to allow for waste.

Bricklaying Terminology

Before explaining the proper procedure required for quality brick structures, let us define some of the more specific terms used to describe the various positions of the brick units and mortar joints in a wall.

- Bull-header. A rowlock brick laid with its longest dimension perpendicular to the face of the wall.
- Bull-stretcher. A rowlock brick laid with its longest dimension parallel to the face of the wall.
- Course. One of the continuous horizontal rows of masonry which, bonded together, form the masonry structure.
- Header. A masonry unit laid flat with its longest dimension perpendicular to the face of the wall. It is generally used to tie two wythes of masonry together.
- Rowlock. A brick laid on its edge (face).
- Soldier. A brick laid on its end so that its longest

dimension is parallel to the vertical axis of the face of the wall.
- Stretcher. A masonry unit laid flat with its longest dimension parallel to the face of the wall.
- Wythe. A continuous vertical 4-inch or greater section or thickness of masonry, as the thickness of masonry separating flues in a chimney.

Types of Bonds The word bond, when used in reference to masonry, may have several different meanings:

- Structural bond is the method by which individual masonry units are interlocked or tied together to act as a single structural unit. Structural bonding of brick and tile walls is accomplished in three ways: (1) by overlapping (interlocking) the masonry units, (2) by the use of metal ties imbedded in connecting joints, (3) by the adhesion of grout to adjacent wythes of masonry.
- Mortar bond is the adhesion of the joint mortar to the masonry units or to the reinforcing steel.
- Pattern bond is the pattern formed by the masonry units and the mortar joints on the face of a wall. The pattern may result from the type of structural bond used or may be purely a decorative one in no way related to the structural bond.
- Running bond is the simplest of the basic pattern bonds and consists of just stretchers. Since no headers are used, metal ties are usually used. Running bond is used largely in cavity wall construction, veneered brick walls, and often

	Quantities of Brick and Mortar Needed for Common Brick Walls Using ½-Inch Joints							
	Wall thickness							
	1 brick		2 bricks		3 bricks		4 bricks	
Wall area sq. ft.	Number of bricks	Mortar cu. ft.	Number of bricks	Mortar cu. ft.	Number of bricks	Mortar cu. ft.	Number of bricks	Mortar cu. ft.
1	6.17	0.08	12.33	0.2	18.49	0.32	24.65	0.44
10	61.7	0.8	123.3	2	184.9	3.2	246.5	4.4
100	617	8	1,233	20	1,849	32	2,465	44
200	1,234	16	2,466	40	3,698	64	4,930	88
300	1,851	24	3,699	60	5,547	96	7,395	132
400	2,468	32	4,932	80	7,396	128	9,860	176
500	3,085	40	6,165	100	9,245	160	12,325	220
600	3,712	48	7,398	120	11,094	192	14,790	264
700	4,319	56	8,631	140	12,943	224	17,253	308
800	4,936	64	9,864	160	14,792	256	19,720	352
900	5,553	72	10,970	180	16,641	288	22,185	396
1,000	6,170	80	12,330	200	18,490	320	24,650	440

Note: When using ⅝-inch joints, multiply quantities by 80 percent. When using ⅜-inch joints, multiply quantities by 120 percent. Mortar estimates include 10 percent waste allowance.

in facing tile walls where the bonding is accomplished by extra width stretcher tile.

- Common or American bond is a variation of running bond with a course of full length-headers at regular intervals. These headers provide structural bonding as well as pattern. Header courses usually appear at every fifth, sixth, or seventh course, depending on structural bonding requirements. In laying out any bond pattern, it is very important that the corners be started correctly. For common bond, a three-quarter brick must start each header course at the corner. Common bond may be varied by using a Flemish header course.
- Flemish bond is made up of alternate stretchers and headers, with the headers in alternate courses centered over the stretchers in the intervening courses. Where headers are not used for structural bonding, they may be obtained by using half-brick, called blind-headers. There are two methods used in starting the corners: the so-called Dutch corner, in which a three-quarter-brick is used to start each course, and the English corner, in which 2-inch or quarter-brick closures must be used.
- English bond is composed of alternate courses of headers and stretchers. The headers are centered on the stretchers and joints between stretchers. The vertical (head) joints between stretchers in all courses line up vertically. Blind headers are used in courses which are not structural bonding courses.
- Block or stack bond is purely a pattern bond. All vertical joints are aligned. Usually this pattern is bonded to the backing with rigid steel ties. When 8-inch thick stretcher units are available, they may be used. In large wall areas and in load-bearing construction, reinforce the wall with steel pencil rods placed in the horizontal mortar joints. Vertical alignment requires dimensionally accurate units, or carefully pre-matched units, for each vertical joint alignment.
- English cross or Dutch bond is a variation of English bond and differs only in that vertical joints between the stretchers in alternate courses do not line up vertically. These joints center on the stretchers themselves in the courses above and below.

Laying the Brick

Bricks are laid on a foundation, a footing, or a subwall—which must be smooth and level. The first step is to lay out, without mortar, the first layer of bricks on the foundation, starting at one corner and progressing around the foundation's perimeter. This procedure is called chasing out the bond and permits planning the job so that the bricks will come out evenly at every corner.

Before the first layer is laid out, wet the brick. Unlike cement block, brick must be moist or it will absorb water from mortar, resulting in a poor bond. Wet a pallet or a stack of bricks by soaking them until water begins to run off.

Mark the corners of the wall with nails and snap a chalk line between them. Lay out the brick, leaving space for the mortar. If the bricks do not come out evenly when a corner is reached, the mortar spaces or joints can be readjusted equally so the last brick just fits.

Preparing the Mortar Mortar for brickwork is mixed in the same manner as mortar for concrete block. After the mortar is mixed, it immediately begins to set. It may become stiff and difficult to use. The addition of more water and further mixing tempers the mortar to again make it workable but also weakens its bonding strength. Try to mix mortar in small batches that can be used at once.

Mortar Joints and Pointing The trowel should be held firmly. The thumb should rest on top of the handle and should not encircle it. If you are right-handed, pick up mortar with the left edge of the trowel from the outside of the pile. Pick up enough for one to five bricks, according to the wall space and your skill. A pickup for one brick forms a small windrow along the left edge of the trowel. A pickup for five bricks is a full load.

Holding the trowel with its left edge directly over the centerline of the previous course, tilt the trowel slightly and move it to the right, dropping a windrow of mortar along the wall until the trowel is empty. The remaining mortar is returned to the board. If you are right-handed, work from left to right along the wall.

Mortar projecting beyond the wall line is cut off with the trowel edge and returned to the mortarboard, but enough is retained to butter the left end of the first brick to be laid in the fresh mortar.

With the mortar spread about 1 inch thick for the bed joint, a shallow furrow is made and the brick is pushed into the mortar. If the furrow is too deep, there will be a gap between the mortar and the brick which will reduce the resistance of the wall to water penetration. Mortar for a bed joint should not be spread too far in advance of the laying. A distance of four or five bricks is advisable. The mortar must

The proper way to hold a trowel.

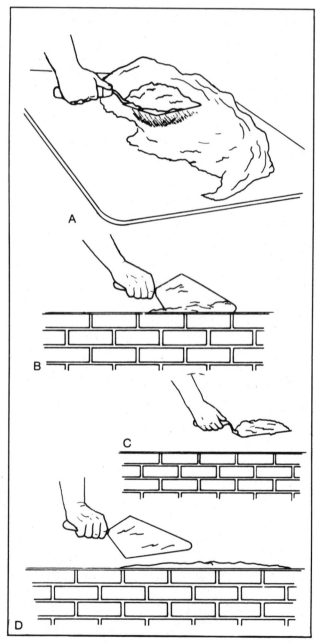

Steps in applying mortar.

be soft and plastic so that the brick can be easily bedded in it.

After the bed joint mortar has been spread, the brick must be laid. The brick is picked up with the thumb on one side of the brick and the fingers on the other. As much mortar as will stick is placed on the end of the brick. The brick is pushed into place so that excess mortar squeezes out at the head joint and at the sides of the wall. The head joint must be completely filled with mortar. After the brick is bedded, excess mortar is cut off and used for the next end joint. The proper position of the brick is determined by the use of a cord.

Cross joints must be completely filled with mortar. The mortar for the bed joint should be spread several brick widths in advance. The mortar is spread over the entire side of the header brick before it is placed in the wall. The brick is then shoved into place so that mortar is forced out at the top of the joint.

Before laying the closure brick of a header course or stretcher course, plenty of mortar should be placed on the sides of the bricks already in place. Mortar should also be spread on both sides of the closure brick to a thickness of about 1 inch. The closure brick should be laid in position without disturbing the brick that is already in place. If any of the adjacent bricks are disturbed, they must be removed and relaid to prevent cracks from forming between the brick and mortar.

There is no hard and fast rule regarding the thickness of the mortar joint. Brick irregularities are taken up in the mortar joint. Mortar joints ¼ inch thick are strongest and should be used when the bricks are regular enough to permit it.

Filling exposed joints with mortar immediately after the wall has been laid is called pointing. Pointing is frequently necessary to fill holes and correct defective mortar joints. The pointing trowel is used for this purpose.

Cutting Brick If a brick is to be cut to exact line, the bolster or brick set should be used. When using these tools, the straight side of the cutting edge should face you and the part of the brick to be saved. One blow of the hammer on the brick set should break the brick. Extremely hard brick will need to be cut roughly with the head of the hammer in such a way that there is enough brick left to be cut accurately with the brick set.

For normal cutting work, the brick hammer should be used. The first step is to cut a line all the way around the brick with light blows of the hammerhead. When the line is complete, a sharp blow to

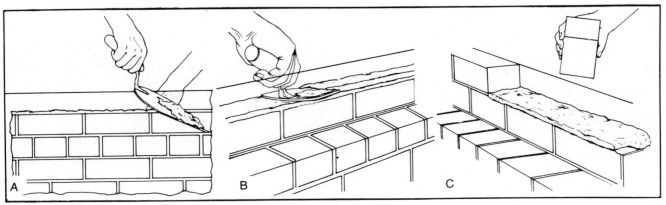

Steps in making a bed joint and furrow.

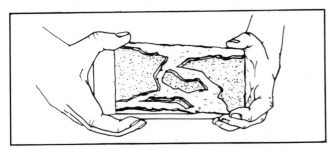

A poorly bonded brick.

STRING

Head joint in a stretcher course.

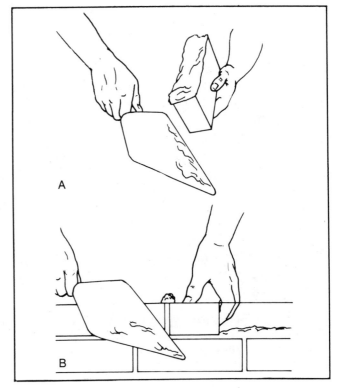

Making cross joints in header courses.

one side of the cutting line will split the brick at the cutting line. Rough places are trimmed using the blade of the hammer or the curved edge of the trowel.

Joint Finishes Exterior surfaces of mortar joints are finished to make the brickwork more waterproof and to improve the appearance. The types of joint finishes are described below:

1. Concave joint. Created through the use of a rounded jointing tool.
2. Vee joint. Created with a Vee-shaped tool.
3. Weathered joint. Made by inclining the joint so that it sheds water readily.
4. Struck joint. Another inclined joint, but the reverse of the weathered joint.
5. Raked joint. Made by removing some of the mortar with a square-edged tool. The struck and raked joints are decorative, but they invite water seepage because water can collect in the joint.
6. Flush joint. This is the simplest joint. It is made by holding the edge of the trowel flat against the brick and striking off the excess mortar, leaving the mortar flush with the face of the brick. A variation is the squeezed joint, made by leaving the excess mortar which is squeezed out as the brick is tapped into place. The most effective joints in terms of their ability to resist moisture are the concave, Vee, and weathered joints.

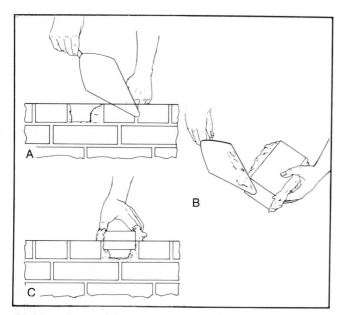

Making closure joints in header courses.

Cleaning Excess mortar should be trimmed as you lay brick. After you tool the joints, loose mortar should be brushed away again with a soft-bristle brush to remove particles left from tooling. Use a wire brush on stubborn spots.

After the mortar sets and cures 7 to 10 days, remaining spots can be removed by washing the wall with soap and water. If necessary, apply a solution of 1 part muriatic acid and 9 parts water. Scrub the stained area and rinse well. Handle muriatic acid solutions with care. Wear safety glasses, long sleeves, and work gloves.

Stone Masonry

In many respects, natural stone masonry and brick masonry are similar. However, stone masonry

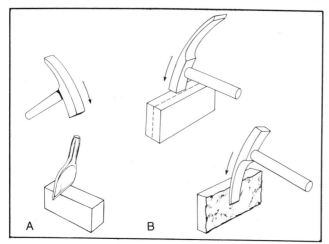

(A) Cutting a brick with a bolster. (B) Cutting a brick with a hammer.

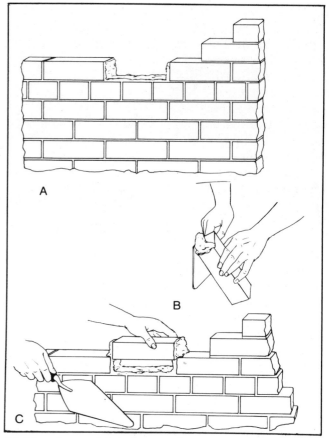

Making closure joints in stretcher courses.

requires more specialized work and considerable care when laying out a job. Stones of many types are used in residential construction, varying from rough, irregular, random-sized native stone to precisely cut limestone laid in a specific pattern.

Stonework falls into two broad categories: rubble stone masonry and ashlar masonry. In rubble stone masonry, the stones are left in their natural, uncut, unshaped state. In ashlar masonry, all surface faces are squared to form a continuous plane. Both rubble and ashlar work may be either coursed or random.

Random rubble is the crudest type of stonework. Little attention is paid to laying the unstratified stones in courses. Each layer contains bonding stones that extend the entire width of the wall. This produces a strong wall. Bed joints are horizontal for stability, but the "builds" or head joints may run in any direction.

Coursed rubble is assembled of roughly squared stones laid in continuous horizontal bed joints.

Ashlar walls are laid up primarily with stratified stones, such as sandstone and limestone, because they are easy to cut and dress. Coursed or ranged ashlar patterns consist of square-cut stone laid in regular courses, much like brick and concrete block. Random or broken ranged ashlar patterns are constructed of different size square-cut or roughly square-cut stones that are not arranged in any specific design. Although ashlar walls are easier to construct, cut and dressed stones are generally more expensive. Most ashlar masonry is made with weaker stratified stone, so the maximum length of each stone should not exceed three times its height. With stronger unstratified stones, such as granite, the maximum length is up to five times the height. The maximum width should be about one-and-one-half times the height in stratified stones to three times the height in unstratified stones.

Materials for Use in Stone Masonry Common stones used in stonework include limestone, sandstone, granite, and slate. Unsquared stones may be found on the building site itself or on nearby land. It is generally possible to obtain permission to collect these stones. Stream beds and the banks of streams are excellent sources of fieldstone in many regions.

Free stone is often available in abandoned quarries, fields, crumbling barn or stone foundation walls, or from building demolition contractors or road excavation crews.

Commercial quarries and stone yards sell stone in three grades: dressed, semidressed, and undressed. The undressed grade consists of stones which are uncut and unfinished and are generally the cheapest grade. Semidressed stone has squared-off corners, but the edges are roughly parallel. The stone is not cut to any particular size. Dressed stone is cut to a specific size and sometimes can be ordered custom cut. This is the most expensive grade, but it is laid in courses much like building brick or solid cement block.

The larger the stone, the more quickly the work goes. Large stones fill most of the area, and the smaller ones fill in between to conserve mortar. Avoid stones that you cannot lift comfortably by yourself. Comfortably means lifting for a sustained period.

Random rubble masonry.

Most stone is sold by the ton, but some dealers sell it by the cubic yard. Required tonnage differs from undressed to dressed stone. One ton of rubble or undressed will generally cover from 25 to 45 square feet of wall, with an average thickness of 1 foot. One ton of dressed stone will cover approximately 50 to 60 square feet of wall surface, with an average thickness of 6 inches.

Most stone yard or quarry personnel can estimate how much stone is needed for a given project. Calculate the cubic footage of the area to be covered with stone and take your figures with you. Stone should match in color and texture with a good mix of sizes. If you are ordering dressed stones, specify the thicknesses you want and designate minimum and maximum lengths. Larger stones can probably be custom cut if necessary.

Stone masonry mortar should be composed of portland cement, lime and sand. See the following table for proper ratios.

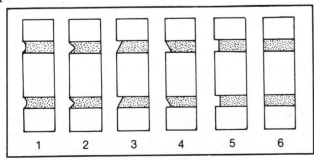

The most popular brick mortar joints.

Recommended Mortar Mixes for Stonework			
Proportions by Volume			
Type of Service	Cement	Hydrated Lime	Mortar sand in damp, loose condition
For ordinary service.	1—masonry cement* or	—	2¼ to 3
	1—portland cement.	½ to 1¼	4½ to 6
Subject to extremely heavy loads, violent winds, earthquakes, or severe frost action. Isolated piers.	1—masonry cement* plus 1—portland cement or	—	4½ to 6
	1—portland cement.	0 to ¼	2¼ to 3
*ASTM Specification C 91 Type II.			

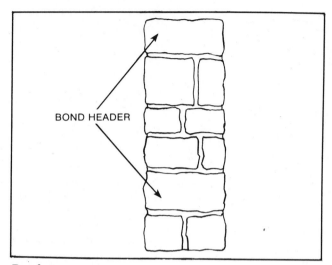

Bond stones.

Coursed rubble masonry.

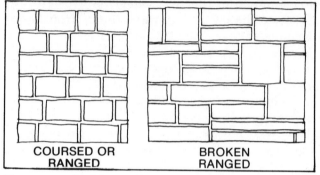

Two types of ashlar walls.

Three grades of stones.

Laying Stone with Mortar The wall proper should be laid on a footing of large stones, each as long as the footing is wide. Some general rules for laying stones are:

Each stone is laid on its broadest face. If appearance is a factor, larger stones should be laid in the lower courses. The size of stones should gradually diminish toward the top of the wall. This is not possible with most ashlar designs, however, because of the decorative patterns.

Masonry stone is classified as absorbent or nonabsorbent. Absorbent stone must be wetted before placing, to prevent absorption of water and a subsequent weakening of the mortar. The nonabsorbent type requires no wetting.

Stones should be selected and placed to make the spaces between stones as small as possible. If large spaces are unavoidable, they must be filled with small stones embedded in mortar.

If a stone must be moved after it has been placed, it must be lifted out entirely and reset.

The thickness of bed joints depends on the size and type of the stone. Mortar must be thick enough to fill all the spaces between stones.

Head joints are not made by buttering, as with brick, but by slushing with mortar and, if necessary, filling with small stones after three or four stones have been laid on bed joints.

Bond stone should be laid in every 6 to 10 square feet of wall.

Joints and Pointing There are two classifications for horizontal joints between stones, bed joints or simply beds. Vertical joints are called head joints or builds.

Rubble masonry joints are neither constant in direction nor uniform in thickness, because they are used to fill the spaces between irregularly shaped stones.

Ashlar or cut-stone masonry joints do not exceed ½ inch in thickness because the stones are accurately dressed. A joint thickness of ¼ inch is used for ashlar facing and a thickness of ⅛ inch for interior stonework.

Where joints of ashlar masonry are raked out to a depth of about ¾ inch, pointing is necessary. Sometimes a special mortar may be used to make a tighter and more attractive joint. Pointing should not be performed while the wall is being constructed, but must be done only after the mortar has fully set and the wall has received its full load.

Rubble masonry does not require pointing. Joints are considered finished when the stones are fully set. Joints of rubble masonry can be made flush with the surface of the stones. This process is also done after the mortar has set.

Cutting and Splitting Stone To cut a relatively flat stone, score the cutting line using a straightedge and mark the line with chalk or with the corner of a stonecutter's chisel. Position the line over a board, and cut a groove ¼ inch deep with a chisel and sledgehammer. A circular saw with a diamond or abrasive blade will effectively score the rock. Make repeated passes, lowering the blade each time until the groove is deep enough. When the groove has been cut or chiseled on one side, turn the slab over and repeat scoring and grooving.

Position the stone so that the line overhangs the board 1 inch or so, then hit the waste piece gently with the sledge until it drops off. The break may not be even. To trim the stone, hold the chisel at an angle to the uneven edge and rap with the sledge until sufficient rock chips off.

Finding rock grain and learning how to utilize it are the most important skills in splitting round stones into flat slabs. The grain of a stone looks like a series of light, parallel lines that are barely discernible in the hard stones like granite, but easier to identify in limestone or soft sandstone. Splits made along the grain will produce relatively smooth surfaces.

To cut a moderate-sized stone to shape, groove deeply on all sides with a broad-blade chisel. To cut along the grain, lay the broad-blade chisel along the grain line and tap with the hammer several times. Follow the grain around, each time making the groove deeper, until the stone splits. To cut across the grain, groove around all the sides with a broad-blade chisel. After grooving deeply, tap lightly on the outer end of the cutoff until it breaks free. Small points can be knocked off by a light blow directed outward from the center of the stone with a hammer. This flaking can be used to make straight-line cuts or curves.

To cut a large stone, it is generally easiest to use a plug-and-feathers. Bore a hole about one-third the stone's depth along the grain line where you want the stone to split.

When boring a plug-and-feather hole with an electric drill, use a carbide-tipped masonry bit with the drill set on slow speed. If you have a single-speed drill, attach a speed reducer. Use moderate pressure in short intervals, withdrawing the drill often to clean the hole and avoid overheating the motor.

Insert the plug-and-feathers and strike the plug with the sledgehammer until the stone splits. If the stone is really large, you may need several holes with several plug-and-feathers.

Cleaning Stone Unlike other masonry, most stone should not be washed down with any acid or caustic cleaner. These substances cause the surface of the stone to disintegrate or stain. To clean granite and bluestone, use a nonabrasive, acid-free detergent. Slate can be washed with hot water and a mild detergent. On the other hand, limestone and sandstone can be scrubbed with a fiber brush (avoid synthetics) and water. Detergents can mar the surface of soft stones. Interior soft stones can be cleaned with a dry, soft-bristle brush only. To remove stains or burn marks from limestone or sandstone, rub the area with a medium-grade abrasive paper, then clean.

Structural Walls of Brick, Concrete Block, and Stone

Previous chapters explained the fundamentals of planning and assembling a single brick or concrete block wall. This chapter will discuss how to construct building walls that are two masonry units thick, and other basic wall building techniques. Included are details on building safe door and window openings, strengthening a wall with pilasters, constructing intersecting walls, and waterproofing a masonry wall.

Structural masonry walls are either veneered, solid, or cavity. The most common brick wall used in home construction is a veneered wall consisting of a single layer of facing bricks laid up outside the wall sheathing material. The structural wall itself is framed in the conventional manner with wood studs. The bricks act as an exterior finishing material attached to the frame wall by metal strips called ties. Ties are nailed to the sheathing and embedded in the mortar between bricks.

The second type of brick wall used in houses is a solid masonry wall. There is no wood frame behind the bricks. The bricks, and usually a backup of hollow masonry units or bricks, provide both the enclosure and structural system. Furring strips can be fastened to the masonry units, and the interior can be finished with wall covering material applied to the furring strips. Waterproofing is very important since both the bricks and mortar will readily absorb moisture. If water seeps through the wall, the interior finish will be damaged unless proper precautions are taken. A coating or two of hot tar applied to the inside surface of the solid brick wall provides sufficient protection.

The third type of brick wall is known as a cavity wall because a space is left between an outer and an inner width of bricks. The space, approximately 2 inches wide, may be filled with insulation or be left as a dead air space. The dead air space blocks the passage of water through the wall and also provides some insulation. Cavity wall construction is often used for exposed brick interior walls.

The three types of masonry walls can also be constructed with concrete blocks or stones substituted for the bricks.

Brick Structural Walls

The principal factors governing brick masonry wall strength are:
- Brick strength
- Mortar strength and elasticity
- Workmanship of the bricklayer
- Brick uniformity
- Laying method used

The strength of an individual brick varies widely, depending upon the material and manufacturing method. Bricks with compressive strengths as low as 1,600 pounds per square inch have been made, whereas some well-burned bricks have compressive strengths exceeding 15,000 pounds per square inch.

The strength of portland cement-lime mortar is normally higher than the strength of the bricks, so brick masonry laid in cement-lime mortar is stronger than the strength of the individual brick. The use of plain lime mortar reduces the load-carrying capacity of a wall or column to less than half the load-carrying capacity of the same type construction built with portland cement-lime mortar.

Footing The first step in brick wall construction is building a footing. While most footings are concrete, bricks can be used. The required footing width and thickness are basically the same as those for concrete. Every footing should be below the frost line to prevent the foundation from settling. Although brickwork footings are satisfactory, footings are usually concrete, leveled on top to receive the brick or stone foundation wall.

As soon as the subgrade is prepared for a brick footing or foundation, place a bed of mortar about 1 inch thick on the subgrade to take up all irregularities. The first course of the foundation is laid on this bed of mortar. Other courses are laid on the first course.

The construction method for a column footing is the same as for the wall footing.

Solid 8-Inch Common Bond Brick Wall For a wall of a specific length, make a slight adjustment in the width of head joints so that an even number of bricks will make up the length. First, lay the

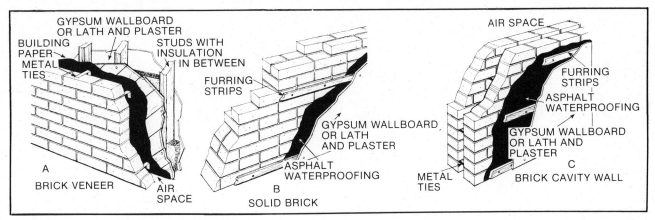

Three types of masonry structural walls.

bricks on the foundation without mortar. The distance between the bricks is equal to the thickness of the head mortar joints. Corners are erected first. This is called laying of leads. You will use these leads as a guide in laying the remainder of the wall.

The primary phase in laying a corner lead is shown in the first step of the illustration on page 50. Two three-quarter closures are cut, and a 1-inch thick mortar bed is laid on the foundation. The three-quarter closure A in the second step is pressed down into the mortar bed until the bed joint is ½ inch thick. Next, mortar is placed on the end of another three-quarter closure B, and a head joint is formed as previously described. The head joint between the two three-quarter closures should also be ½ inch thick. Excess mortar squeezed from the

joints is cut off. The level of the two three-quarter closures should be checked by means of a plumb rule placed in the positions indicated by the dashed lines in the second step of the illustration. The edges of these closure bricks must be even with the outside face of the foundation. Next, mortar is spread on the side of brick C, and it is laid as shown in the third step. Its level is checked using the plumb rule in the position indicated by the dashed lines in the third step. Its end must also be even with the outside face of the foundation. Brick D is laid and its level and position is checked. When brick D is in the proper position, the quarter closures E and F should be cut and placed. Excess mortar should be removed, and the tops of these quarter closures should be checked to see that they are at the same level as the top surfaces of the surrounding bricks.

In the fourth step, brick G is pushed into position after mortar has been spread on its face. Excess mortar should be removed. Bricks H, I, J, and K are laid in the same manner. The level of the bricks is

FIRST AND SECOND COURSE

THIRD COURSE

FOURTH COURSE

FOOTING AND FOUNDATION COMPLETED

Brick wall footing.

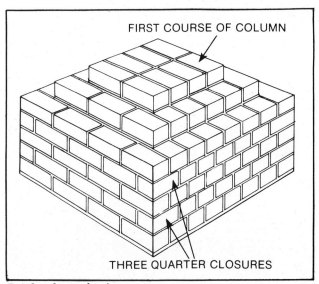

FIRST COURSE OF COLUMN

THREE QUARTER CLOSURES

Brick column footing.

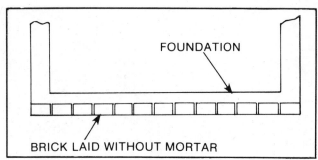

Determination of the vertical brick joints and the number of bricks in one course.

checked by placing the plumb rule in the positions indicated in the fourth step. All brick ends must be flush with the surface of the foundation. Bricks L, M, N, O, and P are then laid in the same manner. The number of leader bricks that must be laid in the first course of the corner lead can be determined from the fifth step. It will be noted that six header bricks are required on each side of the three-quarter closures A and B.

The second course, a stretcher course, is now laid. A 1-inch thick layer of mortar should be spread over the first course and a shallow furrow made in the mortar bed as shown in the next illustration. Brick A is laid in the mortar bed and shoved down until the mortar joint is ½ inch thick. Brick B is shoved into place after mortar has been spread on its end. Excess mortar is removed, and the joint is checked for thickness. Bricks C, D, E, F, and G are laid in the same manner and checked to make sure they are plumb and level. Level is checked by placing the plumb rule in the position indicated in step two. The bricks are plumbed by using the plumb rule in a vertical position. This should be done in several places. As may be determined from step three, seven bricks are required for the second course. The remaining bricks in the corner lead are laid in the manner described for the bricks in the second course.

Since the wall between the leads is laid using the leads as a guide, the level of the courses in the lead must be checked continually and carefully. After the first few courses, the lead is plumbed. If the brickwork is not plumb, bricks must be moved in or out until the lead is accurately plumb. It is not good practice to move a brick very much once it is laid in mortar; therefore, care is taken to place the brick accurately. Joints are tooled or finished before the mortar has set.

A corner lead at the opposite end of the wall is built in the same manner. The top of the second course in one corner lead must be the same height above the foundation as the second course in the other corner lead. A long 2 x 2-inch pole can be used

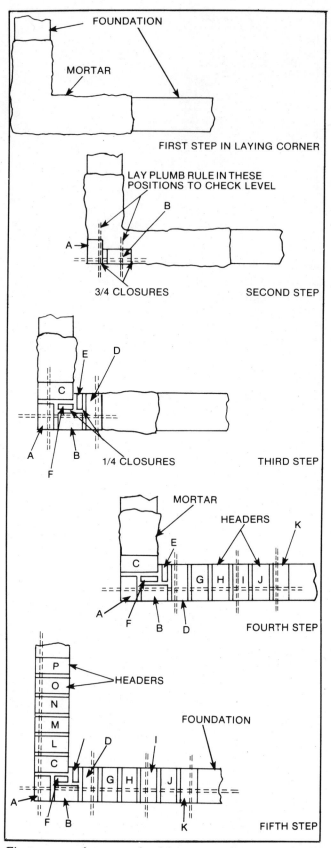

First course of a corner lead for an 8-inch common bond brick wall.

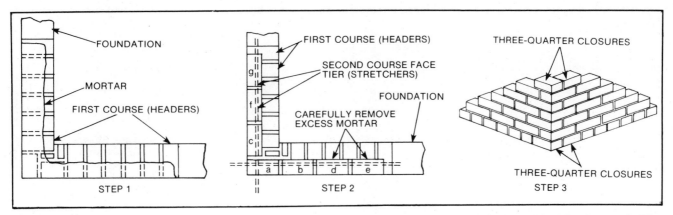

Second course of a corner lead for an 8-inch common bond brick wall.

to mark off the heights of the different courses above the foundation. This pole can be used to check the course height in the corner leads. Laying leads should be closely supervised. Only skilled builders should attempt this work.

With the corner leads at each end of the wall completed, the face tier of bricks for the wall between the leads is laid. It is necessary to use a line. Knots are made in each end of the line to hold it within the slot of the line block. The line is made taut by hooking one of the line blocks to each end of the wall. The line is positioned $^1/_{16}$-inch outside the wall face, level with the top of the bricks.

With the line in place, the first or header course is laid in place between the two corner leads. The brick is pushed into position so that its top edge is $^1/_{16}$-inch behind the line. Do not crowd the line. If the corner leads are accurately built, the entire wall will be level and plumb. Check the wall between leads with the level at several points. For the next course, the line is moved to the top of the next mortar joint. Finish the face joints before the mortar hardens.

When the face tier between the leads has been laid, normally six courses, the backup tier is laid. The backup bricks for the corner leads are laid first and the remaining bricks are laid afterwards. The line need not be used for the backup bricks in an 8-inch wall. When the backup courses have been laid, the second header course is laid.

The wall for the entire building is built to a specific height, at which time corner leads are continued six more courses. The wall between the leads is constructed as before, and the entire procedure repeated until the wall has been completed to the required height.

Solid 12-Inch Common Bond Brick Wall   The 12-inch thick common bond brick wall requires construction similar to that for the 8-inch wall, with the exception that a third tier of bricks is used. The header course is laid first, and the corner leads are built. Two tiers of backing bricks are required instead of one. Two header courses are required, and they overlap. A line should be used for the inside tier of backing bricks for a 12-inch wall.

Window and Door Openings   Openings for windows are left as the bricklaying proceeds. All windowsills should rest on the same course. When the distance from the foundation to the bottom of the window sill is known, the number of courses required to bring the wall up to sill height can be determined. If the sill is to be 4 feet, 4¼ inches above the foundation and ½-inch mortar joints are to be used, 19 courses will be required. Each brick plus one mortar joint is 2¼ + ½ = 2¾ inches. One course is thus 2¾ inches high. Four feet, 4¼ inches divided by 2¾ is 19, the number of courses required.

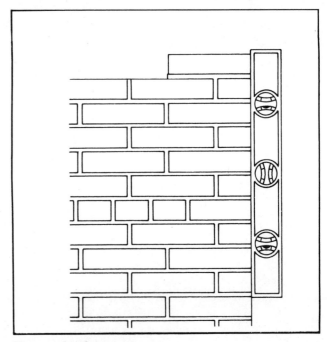

How to plumb a corner.

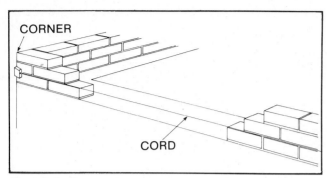

Use of the line.

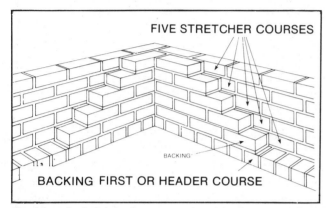

Backing the brick at the corner of an 8-inch common bond brick wall.

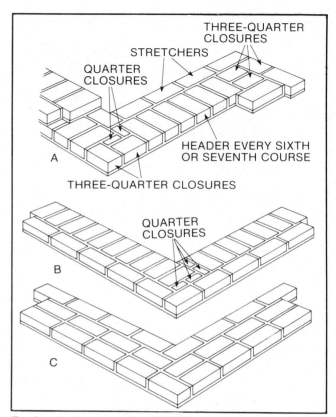

Twelve-inch common bond wall.

With the bricks laid to sill height, the rowlock sill course is laid. The rowlock course is pitched downward normally taking up a vertical space equal to two regular courses of brick. The exterior surface of the joints between the bricks in the rowlock course must be carefully finished to make it watertight.

The window frame is placed on the rowlock sill as soon as the mortar has set and is temporarily braced until the brickwork has been laid to about one-third the height of the frame. These braces are not removed for several days so that the wall above the window frame will set properly. The remainder of the wall is laid so that the top of the bricks in the course level with the top of the window frame is not more than ¼ inch above the frame. Wall height can be adjusted through the addition or deletion of courses, expansion or reduction of the joint, or a combination of both. The corner leads should be laid after the heights of each course at the window is determined.

Lintels are placed above windows and doors to support the weight of the wall above them. They rest on the brick course that is level or approximately level with the frame head, and are firmly bedded in mortar at the sides. Lintels are made of steel, precast reinforced concrete beams, or wood. The thickness of the angle for a two-angle lintel should be

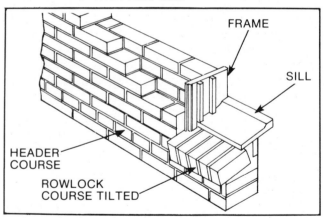

Construction at a window opening.

¼ inch. This makes it possible for the two-angle legs that project into the brick to fit exactly in the ½-inch joint between the face and backup ties of an 8-inch wall.

Space between the window frame and lintel is closed with blocking and weather-stripped with bituminous materials. The wall is then continued above the window after the lintel is placed.

The procedure that was used for laying brick around a window opening, including placement of the lintel, can also be used for laying brick around a door opening. The frames of doors and windows

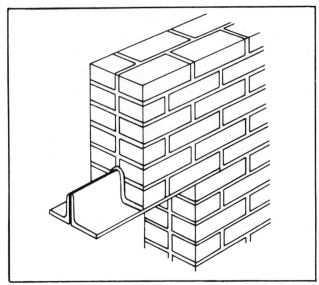

Lintel for an 8-inch wall.

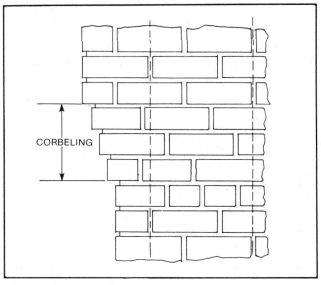

Corbeled brick wall.

are shimmed with wedges to square them. Then, the expansion anchors or lead shields are installed to provide a means for securing the frames.

Corbeling consists of courses of brick set out beyond the face of the wall in order to form a self-supporting projection. The portion of a chimney that is exposed is frequently corbeled out and increased in thickness to improve its weather resistance. Headers should be used as much as possible in corbeling. Use various-sized bats. The first projecting course may be a stretcher course if necessary. No course should extend out more than 2

inches beyond the course below it, and the total projection of the corbeling should not be more than the thickness of the wall.

Corbeling must be done carefully for the construction to have maximum stength. All mortar joints should be completely filled with mortar.

Brick Cavity Walls There are two common types of cavity walls: the standard cavity and the rowlock wall. With the standard brick cavity wall, headers are not required because the two tiers of bricks are held together by means of metal ties installed every sixth course on 24-inch centers. To prevent waterflow to the inside tier, ties must be angled downward from the inside tier to the outside tier.

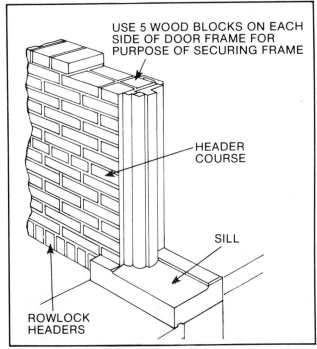

USE 5 WOOD BLOCKS ON EACH SIDE OF DOOR FRAME FOR PURPOSE OF SECURING FRAME

HEADER COURSE

SILL

ROWLOCK HEADERS

Construction at a door opening.

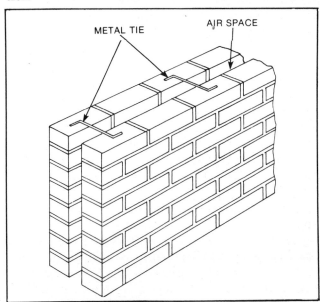

METAL TIE   AIR SPACE

Details for a cavity wall.

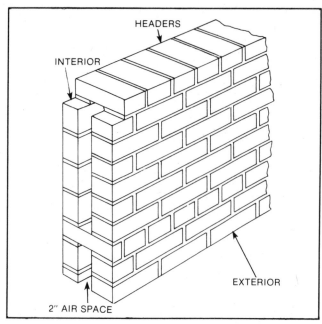

Details of a rowlock back wall.

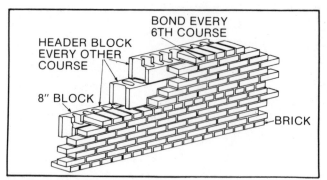

Details of brick-concrete block veneer wall.

The 2-inch cavity between the two tiers of bricks provides a space where water that penetrates the outside tier may flow without passing through to the inside of the wall. The bottom of the cavity is above gound level and is drained by weep holes in the vertical joints between bricks in the first course of the outer tier. The holes should be spaced about 24-inches apart. The air space also gives the wall better heat- and sound-insulating properties.

The face tier of the common rowlock wall has the same appearance as a common bond wall with a full header course every seventh course. The backing tier is laid with the bricks on edge. The face tier and backing tier are tied together by a header course. A 2-inch space is provided between the two tiers of brick.

An all-rowlock wall is constructed with bricks in both the face and backing tier laid on edge with a header course installed at every fourth course. A rowlock wall is not as watertight as a cavity wall.

Partition walls in most masonry projects carry very little load and can be made using one tier of bricks 4 inches thick, laid without headers.

Bricks are laid in cavity walls and partition walls according to the procedure given for making bed joints, head joints, cross joints, and closures. The line is used in the same way as for a common bond wall. Corner leads for these walls are erected first, and the wall between is built afterward.

Veneered Brick Walls The use of one wythe of bricks as a veneer on frame or masonry walls gives the appearance of a solid brick wall while providing economy in construction and better insulation. Brick veneer is simply an addition to the regular frame structure of conventional wood construction. The bricks are tied to the frame with 22-gauge corrugated steel strips spaced 16 or 24 inches on center. There should be a 1-inch air space between the veneered wall and the sheathing. Brick veneer can also be placed on a masonry block wall. Brick veneer on a frame wall consists of stretcher bonding throughout, whereas when used with a masonry wall, it can be bonded by headers at specified intervals.

Brick veneer is frequently used in remodeling and renovating older homes. Old siding does not have to be removed when using a masonry veneer. You must consider how far the present roof extends beyond the sides of the house. With a large overhang, the new veneer can be brought right up to the roof without difficulty. If the overhang is not large enough, the veneer can be brought up to first-floor height, and the gable refinished in wide siding or prefinished panels. If your house does not have gables at the front side, the veneer can be carried to the eaves. Decide these matters in advance.

As with most masonry projects, adding brick veneer begins with digging. Follow the present foundation down to the footing. If it extends out 6 inches

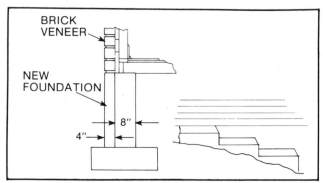

The brick veneer must rest on a firm footing bonded to the house foundation (left). If a house is built on a slope, use step-type footing beneath the veneer (right).

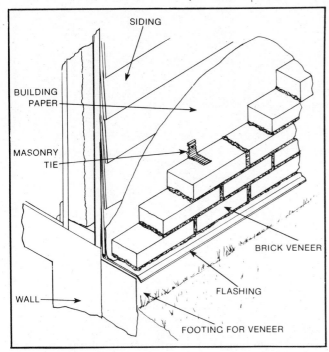

Typical brick veneer construction used over old siding when remodeling.

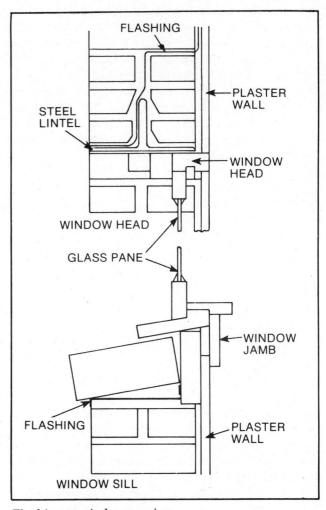

Flashing at window opening.

or more, your new foundation may rest on that. If the footing is less, it must be extended by adding more concrete. To save materials and money, use concrete blocks or poured concrete to within a few inches of ground level and start the veneer at that point. The new foundation must be tied to the old. To ensure a good bond wash the old wall and coat it with grout, a loose mixture of coarse sand and cement.

Once the foundation is ready, you can lay the brick veneer wall. When you reach a window opening, follow the line of the trim, bring the bricks up under the sill, and cement all joints between wood and bricks. The window frames and sill are set in rather than protruding from the house. Across the tops of doors and windows, if the brickwork extends that high, it is necessary to use an L-shaped steel lintel long enough to have a bearing surface of at least 6 inches at each side of the opening. This is set back ½ inch from the outside, with the next course laid on the lower flange of the lintel. Building supply dealers can furnish you with the right-size lintel for your project.

Metal ties, nailed to the frame wall studs, should extend into the mortar joints of every other course. These ties are made of corrugated metal and hold the veneer wall tightly to the old house wall.

If you find one brick out of line, do not try to reset it if the mortar has begun to set. Remove both brick and setting mortar and reset the brick with fresh mortar.

Watertight Walls Resistance to weathering depends on the wall's resistance to water penetration, because freezing/thawing is the only type of weathering that affects brick masonry. Quality workmanship will resist rain penetration during a storm lasting 24 hours accompanied by a 50-mile-per-hour wind. It is unreasonable to expect the type of workmanship that will allow no water penetration. It is advisable to provide some means of taking care of moisture after it has penetrated the brick wall. Properly designed flashing and cavity walls are two ways of handling this moisture.

Flashing is an impervious membrane installed in the mortar joints at certain spots in brick masonry. Flashings exclude water and collect moisture that does penetrate the masonry, directing it to the outside. Flashing is installed at the head and sill of window openings and, in some buildings, where walls and roof meet. Where chimneys pass through the roof, flashing extends entirely through the chimney wall folded up 1 inch against the flue lining.

The edges of the window flashing are turned up to prevent drainage into the wall. Drainage for the wall above the flashing is provided by placing ¼-inch cotton-rope drainage wicks or dowels in the mortar joint just above the flashing membrane spaced 18 inches apart.

Copper, lead, aluminum, and bituminous roofing paper may be used for the flashing membrane. Copper is generally preferred, but will stain the masonry when it weathers. If staining is undesirable, use lead-coated copper paper. Bituminous roofing papers are the cheapest, but are not durable. The cost of replacing worn bituminous paper is many times the cost of installing high-quality flashing initially. Corrugated copper flashing sheets are available that produce a good bond with the mortar. These sheets have interlocking watertight joints at points of overlap.

The flashing is firmly pushed down into a ½-inch thick bed of mortar spread on top of the bricks. The sill or bricks that go on top of the flashing are pushed into a ½-inch thick mortar bed spread on the flashing.

Details for the proper installation of flashing at the head and at the sill of a window are shown in the illustration. At the steel lintel, the flashing goes in under the face tier of bricks, then in back of the face tier, and finally over the top of the lintel.

Water that passes through brick walls usually enters through cracks between the mortar and bricks. Cracks are formed because the bond is poor. They are more likely to occur in head joints. If the position of the bricks is changed after the mortar begins to set, the bond will be destroyed, and a crack will result. Mortar shrinkage is frequently responsible for cracks.

Cracks can be minimized if the exterior face of all the mortar joints is tooled to a concave finish. All head joints and bed joints must be completely filled with mortar to prevent water penetration.

A procedure effective in producing a waterproof wall is plastering the backs of the bricks in the face

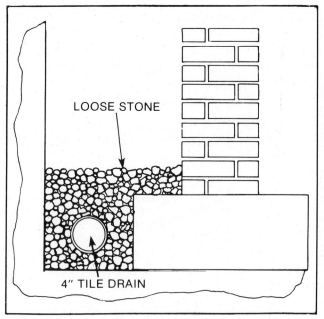

The drain around a typical brick foundation.

tier with not less than ⅜ inch of rich cement mortar before the backing bricks are laid. This is called parging or back plastering. All joints on the back of the face tier of bricks must be cut flush for parging.

Membrane waterproofing, installed as specified for concrete walls, should be used if the wall is subject to considerable water pressure. A properly installed membrane adjusts to shrinkage or settlement without cracking. If the wall is to be subjected

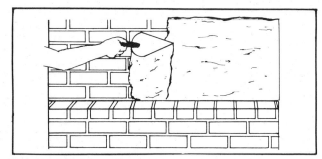

Parging a brick wall.

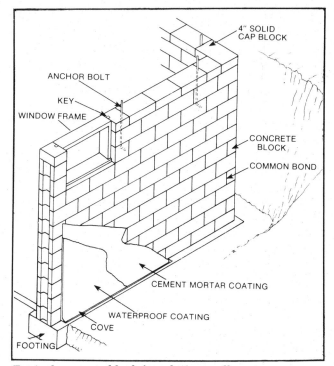

Typical concrete block foundation wall.

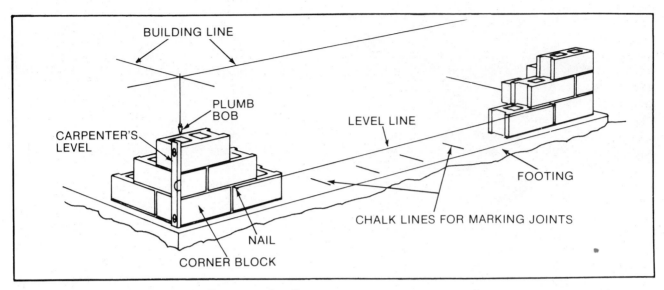

Laying out a typical concrete block structural wall.

to considerable ground water, or if the surrounding soil is impervious, tile drains should be constructed around the base of the wall.

For a foundation wall below ground level, two coats of tar applied to the outside surface of the bricks will provide satisfactory protection.

Watertightness of brick walls above ground level is improved by applying transparent, waterproof paints, such as a water solution of sodium silicate. Varnish is also effective. Certain white and colored waterproofing paints are also available.

Portland cement paint generally gives excellent results. The brick wall should be at least 30 days old before the paint is applied. All efflorescence must be removed from the surface to be painted. Surfaces must be damp when the cement paint is applied. A water spray is the best means of wetting the surface. Whitewash or calcimine-type brushes are used to apply the paint.

Concrete Block Structural Walls

Unlike brick structural walls which are either veneered or contain two courses, most concrete block structural walls are made of single thicknesses and built as described earlier.

Foundation Walls   Foundation block sizes vary somewhat by location and soil conditions. A common minimum thickness for foundation walls in a single-story frame house is 8 inches for hollow concrete blocks.

Concrete block foundation walls require no formwork. Block courses start at the footing and are laid with ⅜-inch mortar joints, in a common bond pat-

tern. Joints are tooled smooth to resist water seepage. A full bedding of mortar should be used on all contact surfaces of the blocks. When pilasters are required to strengthen a wall, they are built on the interior side of the wall and terminated at the bottom of the beam or girder support. Basement door and window frames should be set with keys for rigidity and to prevent air leakage. Block walls are often capped with 4 inches of solid cap blocks. Anchor bolts for sills extend through the top two rows of blocks and the top cap. They should be anchored with a large plate washer at the bottom. Block openings are filled solidly with mortar or concrete.

Full and half length block form a control joint (top). Raking the mortar from a control joint (bottom).

Paper or felt used for a control joint.

Freshly laid block walls must be protected from freezing. If the mortar freezes before it has set, it results in low adhesion, low strength, and joint failure.

To provide a tight, waterproof joint between the footing and wall, use elastic caulking compound. The wall is waterproofed by applying a coating of cement-mortar over the block with a cove formed at the footing. When the mortar has set, a coating of asphalt or other waterproofing will normally ensure a dry basement.

For added protection where wet soil conditions are prevalent, a waterproof membrane of roofing felt or other material can be mopped over the cement-mortar coating, with shingle-style laps of 4 to 6 inches. Hot tar or asphalt is commonly mopped over the membrane to prevent leaks if minor cracks develop in the blocks or joints.

Like poured concrete piers, concrete block piers should extend at least 12 inches above the groundline. The minimum size for a concrete block pier should be 8 x 16 inches, with a 16 x 24 x 8-inch footing. A solid cap block should be used as a top course. Concrete block piers should be no higher than four times their smallest dimension.

Above-Grade Structural Walls Above-grade concrete block structural walls can be laid on either a concrete slab or on a concrete footing. In either case, the blocks are laid as described earlier, but certain building techniques are necessary when installing an above-grade concrete block wall.

Control Joints Control joints are continuous vertical joints built into concrete masonry walls to control cracking caused by stress. The joints permit slight wall movement without cracking and are laid in mortar just as any other joint. Full- and half-length blocks are used to form a continuous vertical joint. If exposed to the weather or to view, they should be caulked. After the mortar is quite stiff, it should be raked out to a depth of about ¾ inch to provide a recess for the caulking. A thin, flat caulking trowel is used to force the caulking compound into the joint.

Another type of control joint, the Michigan joint, is constructed with building paper or roofing felt inserted in the end core of the blocks and extended the full height of the control joint. The paper or felt prevents the mortar from bonding on one side of the joint. To provide lateral support, metal ties can be laid across the joint in every other horizontal course.

Intersecting Walls Intersecting concrete block bearing walls should not be tied together in a masonry bond except at the corners. One wall should terminate at the face of the other wall with a control joint at the point. Bearing walls are tied together with a metal tie bar ¼ x 1¼ x 28 inches, with 2-inch right angle bends on each end. Tie bars are spaced not over 4 feet apart vertically. Bends at the tie bar ends are embedded in cores filled with mortar or concrete. Pieces of metal lath placed under the cores will support the concrete or mortar filling.

To tie nonbearing block walls to other walls, strips of metal lath or ¼-inch mesh galvanized hardware

Tying an intersecting bearing wall.

Tying an intersecting nonbearing wall.

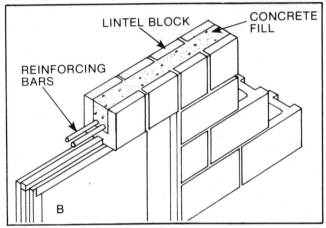

Two styles of lintels.

cloth are placed across the joint between the two walls in alternate courses in the wall. When one wall is constructed first, the metal strips are built into the wall and later tied into the mortar joint of the second wall. Control joints are constructed where the two walls meet.

Lintels The top of door and window openings in masonry construction may be made in two different ways. One is to use a precast concrete lintel which allows the opening to be formed before the door or window frame is set. The other method is to use lintel blocks. Here the frame is set in place, and the block wall is built around it. Lintel blocks are used across the top. Reinforcing bars and concrete are placed in the lintel blocks.

Pilasters Foundation and above-grade walls of both concrete block and brick construction are frequently strengthened by the use of pilasters, which are simply thickened, reinforced sections of the wall. Their purpose is to make the wall strong enough to bear extra heavy vertical loads, such as beam ends, or strong lateral stresses, such as the pressure of wet clay soil.

Frequently, pilaster centers are grouted and reinforced with ⅝-inch steel bars. Grout is prepared by adding extra water to a standard mortar mix to make it thin enough to flow into the spaces and fill them completely.

Walls can also be reinforced with steel bars and grout.

Finishing Concrete Block Walls Concrete blocks can be veneered either with bricks or stones. They can also be painted with oil or latex-based paints. Of course, one of the most popular finishing methods is the use of stucco. Stucco can be applied in the traditional three-coat method, or you can use one of the new acrylic exterior stuccos which requires no measuring or mixing. It is applied directly from the container to create several different textures and is available in five or six pastel colors. Apply the stucco as directed by the manufacturer.

Stone Structural Walls

A structural wall of stone requires considerably more mortar than a brick or concrete block wall. Because of larger joints, as much as one-third of a rubble wall may be mortar.

Solid Stone Structural Walls Stone footings are generally only used for stone or rubble foundations. They are satisfactory if properly laid in strong mortar. The use of stone footings for column supports is not recommended.

The bearing edges of stones for solid structural stone wall construction should be nearly flat. This makes for thinner and stronger joints and conserves mortar. Cobblestone (small, nearly round stones) should be used only as veneer or facing for walls made of other types of masonry.

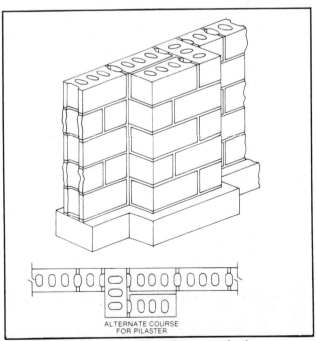

Section of concrete block foundation and pilaster.

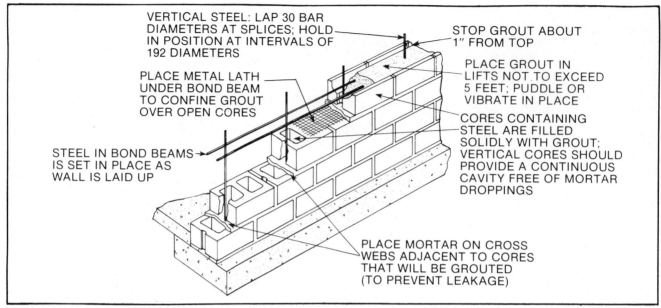

VERTICAL STEEL: LAP 30 BAR DIAMETERS AT SPLICES; HOLD IN POSITION AT INTERVALS OF 192 DIAMETERS

STOP GROUT ABOUT 1" FROM TOP

PLACE METAL LATH UNDER BOND BEAM TO CONFINE GROUT OVER OPEN CORES

PLACE GROUT IN LIFTS NOT TO EXCEED 5 FEET; PUDDLE OR VIBRATE IN PLACE

STEEL IN BOND BEAMS IS SET IN PLACE AS WALL IS LAID UP

CORES CONTAINING STEEL ARE FILLED SOLIDLY WITH GROUT; VERTICAL CORES SHOULD PROVIDE A CONTINUOUS CAVITY FREE OF MORTAR DROPPINGS

PLACE MORTAR ON CROSS WEBS ADJACENT TO CORES THAT WILL BE GROUTED (TO PREVENT LEAKAGE)

Typical single-wythe reinforced masonry block wall.

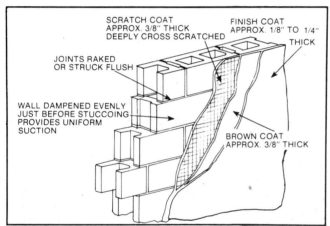

SCRATCH COAT APPROX. 3/8" THICK DEEPLY CROSS SCRATCHED

FINISH COAT APPROX. 1/8" TO 1/4" THICK

JOINTS RAKED OR STRUCK FLUSH

WALL DAMPENED EVENLY JUST BEFORE STUCCOING PROVIDES UNIFORM SUCTION

BROWN COAT APPROX. 3/8" THICK

Stucco on a concrete block wall.

For some types of structural walls, it is wise to taper or slope the wall as it rises. A freestanding wall is designed to slope on the interior wall side about 1 inch for every 2 feet of height. A retaining wall slopes approximately 1 inch for every 1 foot of height. The outer edge of the wall should be kept as plumb as possible, and the vertical joints should overlap at every course. It can also be strengthened transversely by installing headers at right angles to the face of the wall.

All joints should be completely filled with mortar. If the voids are too big for straight mortar, add stone chips to the mix. When laying the wall, be sure all stones fit properly. If they do not, lift the stones out, scrape off all the mortar, and reset in a new bed of mortar.

For above-grade structural walls where appearance is important, rake out the joints on the facing before the mortar sets. The deeper the rake, the better the shadow effect.

Stone Veneer Structural Walls   Most stone veneer is done with cut or dressed stone because the squared surfaces are more easily aligned with the flatness of the backup material. Metal ties must be used to tie the backing material to any type of stone facing. The tabs of the ties that are laid between the courses of bricks or blocks, or nailed to the wood framing, should be bent so that they rest in the mortar between the stones. The stones are then laid against the backup or supporting wall.

Lightweight, natural quarried stone, held in place by special metal clips, is available. This stone can be applied over almost any surface, including wood or plywood, brick, or cement block, on either interior or exterior surfaces. The stone is about 1 inch thick and is grooved at the top and bottom to receive heavy-duty wall anchors secured with special steel tie pins. No special foundation or floor reinforcement is necessary because a steel starter strip sup-

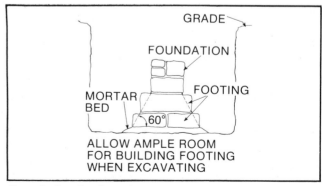

GRADE

FOUNDATION

FOOTING

MORTAR BED

60°

ALLOW AMPLE ROOM FOR BUILDING FOOTING WHEN EXCAVATING

Foundation footing of stone.

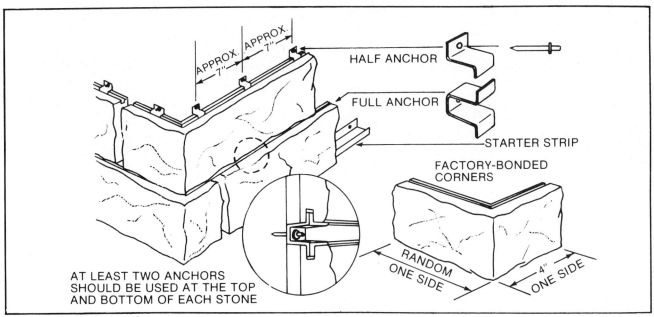

HALF ANCHOR

FULL ANCHOR

STARTER STRIP

FACTORY-BONDED CORNERS

APPROX. 7"

APPROX. 7"

RANDOM ONE SIDE

4" ONE SIDE

AT LEAST TWO ANCHORS SHOULD BE USED AT THE TOP AND BOTTOM OF EACH STONE

Building a stone veneer wall.

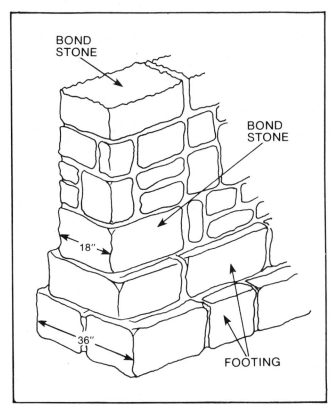

BOND STONE

BOND STONE

18"

36"

FOOTING

Details of a typical stone foundation.

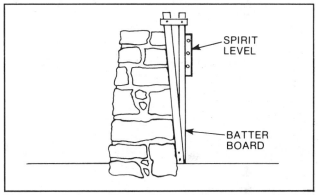

SPIRIT LEVEL

BATTER BOARD

To give a freestanding wall the proper slope, nail three boards together in a triangle and hold it against the face of the wall. The batter board should be set against the sides of the wall; the outside edge should be plumb.

ports the bottom row of stone. A hammer and level are the basic tools needed to install the stone, which is furnished in random lengths from 8 to 24 inches in modular heights of 4 and 8 inches. Factory-bonded corners are also available. After the stones are applied, any mortar can be used to fill the joints.

Applying a stone veneer with steel tie pins.

Driveways, Sidewalks, and Patios

Driveways, sidewalks and patios made of concrete are constructed in basically the same way. The only difference is size and use. The technique of driveway, sidewalk, and patio building applies to most residential concrete construction.

Sidewalks and patios can be constructed of masonry materials other than concrete. Walks and patios of brick and stone are also covered in this chapter. Before explaining construction details, we will look at the design and layout of a driveway, sidewalk, or patio project.

Planning

Driveway Planning   Driveways for single-car garages or carports are usually 10 to 14 feet wide, with a 14-foot minimum width for curving drives. Driveways for two-car garages should be about 16 to 24 feet wide. A driveway should be 36 inches wider than the widest vehicle it will serve. Long driveway approaches to two-car garages may be single-car width but widened near the garage to provide access to both stalls. Instead of one wide slab, a strip driveway may be suitable. It has one concrete strip under each wheel, about 3 to 4 feet wide and 5 feet center to center with space between. The strips in this type of driveway shift more easily with changes in ground conditions than slab driveways.

Driveway thickness depends primarily on the weight of the vehicles that will use it. For passenger cars, 4 inches is sufficient; but if an occasional heavy truck uses the driveway, a thickness of 5 or 6 inches is recommended.

If the garage is considerably above or below street level and is located near the street, driveway grade must be carefully planned. A grade of 14 percent (1¾-inch vertical rise for each running foot) is the maximum recommended. Change in grade should be gradual to avoid scraping the car's bumper or underside.

The driveway should be built with a slight slope toward the street for drainage. A slope of ¼ inch per running foot is recommended. A crown or cross-slope can be used for drainage.

The part of a driveway between the street and public sidewalk is usually controlled by the local municipality. Consult officials when a driveway is built after the street, curbs, and public walks are in place. If the curbs and gutter have not been installed, end the driveway temporarily at the public sidewalk or property line and use an entry of gravel or crushed stone until the curbs and gutter are built. At that time, the drive entrance can be completed to meet local requirements.

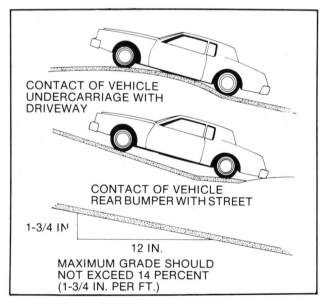

Proper driveway grade is gradual to avoid scraping the car's bumper or underside.

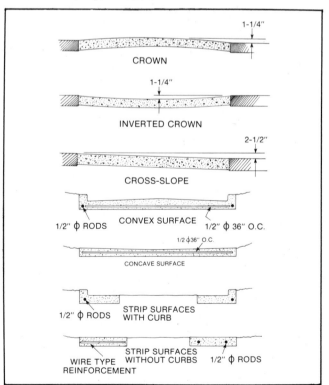

Typical driveway construction sections and how they should be planned to take care of drainage.

If the driveway is built before the public walk, it must meet the proposed sidewalk grade and drop to meet the gutter (if no curb is planned) or the top of any low curb.

Consider using other elements that can make a driveway a beautiful approach to a home rather than just a pathway to the garage.

Sidewalk Planning   Private walks leading to the front entrance of a home should be 3 to 4 feet wide. Service walks connecting to the back entrance may be 2 to 3 feet wide. Private residential walks should be not less than 4 inches thick.

It is customary to slope walks ¼ inch per foot of width for drainage. Walks that abut curbs or buildings should slope toward the curb and away from the building. Where side drainage permits, walks built with a crown or slope from center to edge are desirable. Certain conditions or codes may require that a slope other than ¼ inch per foot be used.

Patio Planning   A carefully planned patio is a valuable extension of the living and entertainment area of any home. Patio planning should consider the view, the climate, traffic flow to the house and kitchen, weather, insect protection, privacy, outdoor cooking, and entertaining.

Location depends on the lot size and house location. If the lot has a beautiful view of the city or surrounding countryside, the patio can be located to take advantage of the view.

Outdoor living should be oriented to the sun and general climate of the region. A south-facing patio is never deserted by the sun. It dries fast after rains and warms quickly in winter months. Patios exposed to the west are likely to be very hot in the afternoon and cool and damp in the morning. Patios facing east are desirable for hot climates because they cool off in the afternoon. North-facing patios never receive direct sun.

The patio should be designed as part of the house. Traffic flow from the house to the patio and location of doors and windows in relation to the patio should be planned. If the floor of the house is above ground level, the patio slab can be built to the level of the house floor. Proper drainage should be maintained on concrete, stone, or brick patios.

If insects are a serious problem, consider enclosing part of the patio with screening. The outdoor living season can be prolonged with a roof for sun or rain protection.

Proximity of neighbors determines the type of privacy screen needed for a patio. Privacy can be obtained with walls, fences, or other visual barriers, such as shrubbery or vines.

Entertaining, cooking, and dining outdoors are among the most enjoyable uses of a patio. The patio should be spacious if considerable entertaining is anticipated. Make the patio larger than the largest room in the house with the food-serving area convenient to the kitchen.

The shape of a patio is limited only by the imagination. Square and rectangular patios are common, but any shape can be built with concrete. A curved or free-form patio can be very attractive, especially when it complements surrounding landscaping.

Building Regulations   Before construction begins on a project, check with the local city or county building department. Most communities require a building permit to ensure that work is done in accordance with the building code. Laws vary with localities. Building permits are especially important for driveways and sidewalks that cross a public way. Some cities will set sidewalk grades when a sidewalk permit is obtained.

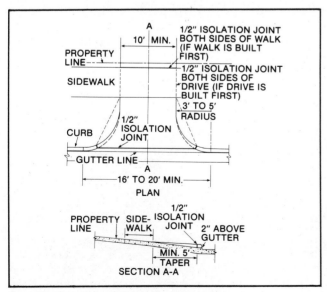

Details of a typical driveway entrance.

This patio was designed level with the floor of the house and convenient to the kitchen and living room.

To prepare a subgrade for a walk, drive, or patio, first remove all sod, roots, and debris.

Use sand or other granular fill to bring the site to a uniform grade.

Preparation for Driveways, Walks, and Patios

The first step in building a walk, drive, or patio is to prepare the subgrade. Serious cracks, slab settlement, and structural failure can usually be traced to a poorly compacted subgrade. The subgrade should be uniform, hard, free from foreign matter, and well drained.

Remove all organic matter such as grass, sod, and roots, and grade the ground. Dig out soft or mucky spots and fill with soil similar to the rest of the subgrade or with granular material, such as sand, gravel, crushed stone, or slag. Compact thoroughly. Loosen and tamp hard spots to provide uniform support.

Granular fills of sand, gravel, crushed stone, or slag are recommended for bringing the site to uniform bearing and final grade. Compact these fills in layers not more than 4 inches thick. Extend the fill at least 1 foot beyond the slab edge to prevent undercutting during rains.

Cover poorly drained subgrades with 4 to 6 inches of granular fill. The bottom of these granular fills must not be lower than the adjacent finished grade, in order to prevent the collection of water under the slab.

Unless fill material can be well compacted, leave the subgrade undisturbed. Undisturbed soil is superior for supporting a concrete slab. Compaction can be done with hand tampers, rollers, or vibratory compactors. For the small job, hand tampers may be used. For large-volume work, mechanical rollers or vibratory compactors are strongly recommended.

A dry spot on the subgrade absorbs more water from the concrete slab than does an adjacent moist spot which may result in dark and light spots in the concrete finish. To prevent this, the subgrade should be uniformly moistened prior to pouring the concrete. There should, however, be no standing water or muddy or soft spots on the subgrade when concrete is being placed.

Formwork Forms for drives, walks, and patios may be made of lumber or metal, and braced by wood or steel stakes. All forms should be straight, free from warping, and of sufficient strength to uniformly resist concrete pressure. Stake and brace the forms firmly to keep them in horizontal and vertical alignment. Setting forms to proper line and grade is normally accomplished with a string line.

Forming and other construction details for a simple driveway or sidewalk and for a typical free-form patio are illustrated.

Set forms so that their tops are level with the string line (left). If there is insufficient room under the string for the form, dig out the subgrade. Attach forms to stakes with nails, which should be driven through the stake and into the form as shown here (center). Hold the form tightly against the stake and level with the string by foot pres- *sure. Double-headed nails are recommended for easy form stripping. For added security, stakes can be braced (right). This is good practice when forming 5- or 6-inch slabs. To prevent the brace from slipping, drive a nail through the brace into the stake.*

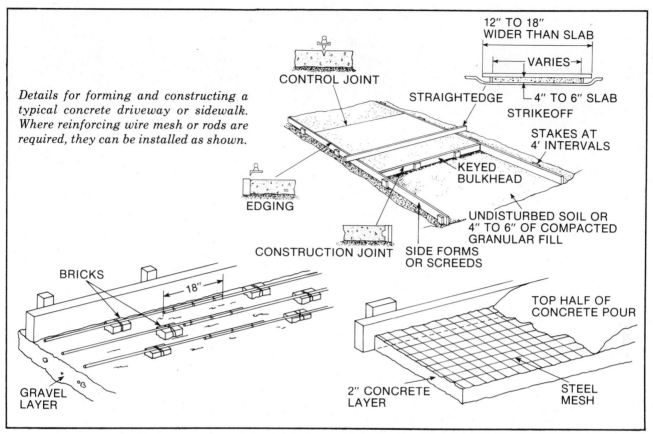

Details for forming and constructing a typical concrete driveway or sidewalk. Where reinforcing wire mesh or rods are required, they can be installed as shown.

CONTROL JOINT

12" TO 18"
WIDER THAN SLAB

VARIES

STRAIGHTEDGE

4" TO 6" SLAB

STRIKEOFF

STAKES AT
4' INTERVALS

EDGING

KEYED
BULKHEAD

CONSTRUCTION JOINT

UNDISTURBED SOIL OR
4" TO 6" OF COMPACTED
GRANULAR FILL

SIDE FORMS
OR SCREEDS

BRICKS

18"

TOP HALF OF
CONCRETE POUR

GRAVEL
LAYER

2" CONCRETE
LAYER

STEEL
MESH

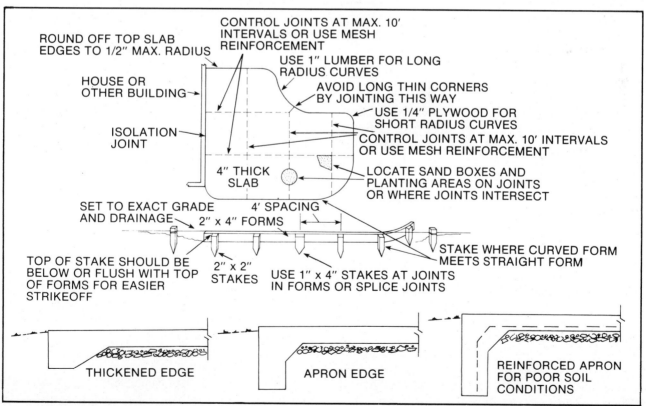

ROUND OFF TOP SLAB
EDGES TO 1/2" MAX. RADIUS

CONTROL JOINTS AT MAX. 10'
INTERVALS OR USE MESH
REINFORCEMENT

USE 1" LUMBER FOR LONG
RADIUS CURVES

HOUSE OR
OTHER BUILDING

AVOID LONG THIN CORNERS
BY JOINTING THIS WAY

USE 1/4" PLYWOOD FOR
SHORT RADIUS CURVES

ISOLATION
JOINT

CONTROL JOINTS AT MAX. 10' INTERVALS
OR USE MESH REINFORCEMENT

4" THICK
SLAB

LOCATE SAND BOXES AND
PLANTING AREAS ON JOINTS
OR WHERE JOINTS INTERSECT

SET TO EXACT GRADE
AND DRAINAGE

4' SPACING

2" x 4" FORMS

STAKE WHERE CURVED FORM
MEETS STRAIGHT FORM

TOP OF STAKE SHOULD BE
BELOW OR FLUSH WITH TOP
OF FORMS FOR EASIER
STRIKEOFF

2" x 2"
STAKES

USE 1" x 4" STAKES AT JOINTS
IN FORMS OR SPLICE JOINTS

THICKENED EDGE

APRON EDGE

REINFORCED APRON
FOR POOR SOIL
CONDITIONS

Details for forming and constructing a typical free-form *patio and the edge construction for concrete patios.*

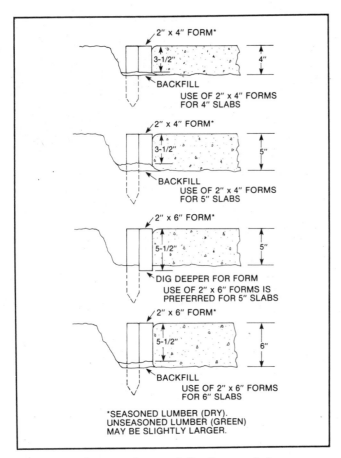

The subgrade must be carefully fine-graded to ensure proper slab thickness when using dressed lumber for formwork.

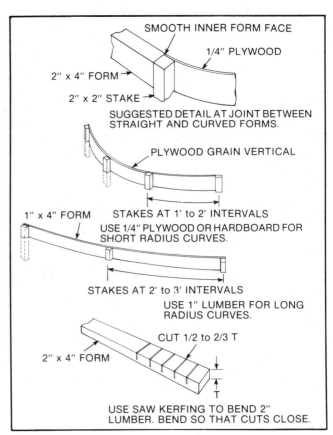

Details for forming horizontal curves.

For a 4-inch-thick slab, 1 x 4 or 2 x 4 lumber may be used. A 5-inch slab can be formed with 2 x 4s, but 2 x 6s are preferable. Slabs of 6-inch thickness require 2 x 6 forms. Lumber used for formwork is generally "dressed," to attain smoothness and uniformity of size. Dressing reduces lumber size; consequently, the 4-inch dimension of a nominal 1 x 4 or 2 x 4 is actually 3½ to 3⁹⁄₁₆ inches, and the 6-inch dimension of a 2 x 6 will be between 5½ and 5⅝ inches. Accordingly, the final grade should be slightly lower than the bottom of the form when using dressed 1 x 4 or 2 x 4 forms for 4-inch slabs. The same applies when using 2 x 4s for 5-inch slabs and 2 x 6s for 6-inch slabs. A little backfilling outside the forms will prevent the concrete from flowing under them.

Wood stakes are made from 1 x 2, 1 x 4, 2 x 2, or 2 x 4 lumber. They may be hand-cut or purchased precut. Space stakes at 4-foot intervals for 2-inch thick formwork. With 1-inch lumber, space the stakes more closely to prevent bulging.

For ease in placing and finishing concrete, drive all stakes slightly below the top of the forms. Wood stakes can be sawed off flush. For easy stripping, use double-headed duplex nails driven through the stake into the form.

Horizontal curves may be formed with 1-inch lumber, ¼- to ½-inch thick plywood, hardboard, or sheet metal. Short-radius curves are easily obtained by bending plywood with the grain in a vertical position. Two-inch thick wood forms may be bent to gentle horizontal curves during staking or to shorter-radius curves by saw kerfing. Wet lumber is easier to bend than dry lumber.

Gentle vertical curves can sometimes be formed by bending a 2 x 4 during staking. When the slope is steeper, short lengths of forming must be used. The curve is laid out with a string line tied to temporary stakes. The line is adjusted up or down on the stakes to give a smooth curve; then, short lengths of forming are set to the string line and securely staked.

To hold forms at proper grade and curvature, set stakes closer on curves than on straight runs.

Wood side forms and divider strips may be left in place permanently for decorative purposes and to serve as control joints. Such forms are usually made of 1 x 4 or 2 x 4 redwood, cypress, or cedar primed with a clear wood sealer. Mask the top surfaces with tape to protect them from abrasion and staining by

The divider strips in this concrete driveway were stained to match the deck, fence, and screen, carrying out the landscape theme.

To preserve their color and protect them from abrasion and staining by concrete, cover the top surfaces of divider strips and outside forms that are to remain in place permanently with masking tape.

concrete. Miter corner joints neatly and join intersecting strips with neat butt joints. Anchor outside forms to the concrete with 16d galvanized nails driven horizontally at 16-inch intervals through the forms at mid-height. Interior divider strips should have nail anchors similarly spaced but driven from alternate sides of the board. Drive all nailheads flush with the forms. Permanent stakes must be driven or cut off 2 inches below the surface of the concrete.

Final Check Before pouring check all forms for trueness to grade and proper slope for drainage. Check the subgrade with a wood template or a string line to ensure correct slab thickness and a smooth subgrade. Dampen forms and subgrade with water. A regular form release agent sprayed on the forms is preferred for volume work.

Pouring Concrete

The most convenient and economical type of concrete for larger jobs is a ready-mix. Ready-mix producers can supply concrete to meet the requirements of most large projects. Methods of determining the amount of concrete needed and how to order it from a ready-mix supplier are given in the first chapter. Small jobs can be mixed by hand or with a mixer.

When ordering your concrete from a supplier, plan to get the truck as close to the point of placement as possible. If the concrete can be poured directly onto the subgrade, try to have a couple helpers available to spread and compact the concrete. Helpers will be needed if the concrete must be transported from the truck to the subgrade by wheelbarrow.

Once the concrete has been poured and compacted to fill the forms, strikeoff and bull-floating or darbying follow immediately. The new poured concrete is struck off with a straightedge, usually a 2 x 4, as described in the first chapter. The stakes are cut off even with the top of the forms to permit continuous movement of the strikeoff board. One to three strikeoff passes should be sufficient. Bull-floating

or darbying follows strikeoff to prevent water from collecting on the surface.

The type of finish you give to your driveway, sidewalk, or patio will depend upon the finishing tool used. A float gives a gritty texture. A stiff-bristle broom can be dragged across the concrete for a rougher texture. The stiffness of the bristles and the pressure applied determine the degree of roughness. Steel-troweling gives a smooth finish that is usually only used on patios.

An edging tool used between the still plastic concrete and the form produces a smooth, rounded edge that will resist breaking and chipping. Control joints must also be provided in order to prevent random cracking of the finished concrete. In driveways and sidewalks, control joints should be spaced at intervals that approximately equal the slab width. Drives and walks wider than about 10 to 12 feet should have a control joint down the center.

Spacing of patio control joints should not exceed 10 feet in either direction. The panels formed by control joints in walks, drives, and patios should be approximately square. Panels with excessive length-to-width ratio are likely to crack. The smaller the panel, the less likelihood of random cracking.

The other type of joint required to control random cracking is the isolation joint, sometimes referred to as an expansion joint. It consists of a premolded ¼- or ½-inch strip of fiber material that extends at least the full depth or length of the slab. Isolation joints are particularly required at points of potential stress, around rigid objects, at intersections of walks and drives, and particularly where a concrete patio abuts the house. Isolation joints are not required at regular intervals in sidewalks and driveways.

Isolation joints should be flush with, or about ¼ inch below, the finished surface. Joints that protrude above the surface are a safety hazard.

After finishing, the concrete must be cured properly. Curing is one of the most inexpensive ways to ensure a long-lasting, satisfactory job.

Curbs There are two ways to form curbs: (1) as an integral part of the driveway, or (2) as a later addition. When adding an addition, drill ½-inch diameter holes 3 inches deep into the drive, spaced 2 feet apart. Insert 6-inch steel reinforcing rods in each hole. The forms for curbs are built by staking a 2 x 8 flat against the driveway edge. Since the inside forms cannot be staked, nail a pair of short 2 x 4s perpendicular to the form, jutting in toward the center of the driveway. Weigh down the 2 x 4 supports with blocks, sandbags, rocks, or bricks. Curbs extend 4 inches above the driveway and are 6 inches wide. Pour and round the inside edge of the curb with a trowel.

Integral curbs are poured at the same time as the driveway. Four-inch concrete slabs with 4-inch high curbs will require 2 x 8 side forms. To mold the rounded inside edge of the curb, cut out the two ends of the strikeoff board to conform to the shape of the curb. When pouring, place a little extra concrete along both sides to fill the curbs completely.

Concrete Patios

There are many ways to dress up a concrete slab. A plain concrete slab can be made more attractive simply by adding wood strips, to break up the monolithic surface and add pattern and contrast in texture. Redwood, highly resistant to rot and decay, is an excellent choice.

In this type of construction, the exterior forms are not removed after the concrete sets, as in ordinary construction. Use redwood for exterior framing. Stake as you would any other frame. Divide the area into boxes with lengths of redwood 2 x 4 stock, notching each piece at the intersections. The cross members are placed on edge and nailed securely to the perimeter framing. The length and the width of the patio should be in multiples of the desired box size.

Use care when pouring and working the concrete to avoid having bits of it dry over the strips. When the mix begins to set, scrub the wood surfaces thoroughly. Final appearance can be improved by applying a redwood sealer/stain before the masonry is poured.

To give a patio the appearance of wood blocks, oil the face of textured plywood and, as the cement sets, press in the oiled surface, withdrawing it immediately. The grain pattern will be imprinted each time. Keep the surface well-oiled to prevent sticking.

Another ingenious way to decorate a patio slab is to cut a stylized outline from heavy building paper or stiff cardboard and press the outline evenly into the wet cement. Let the mix firmly set before removing the pattern.

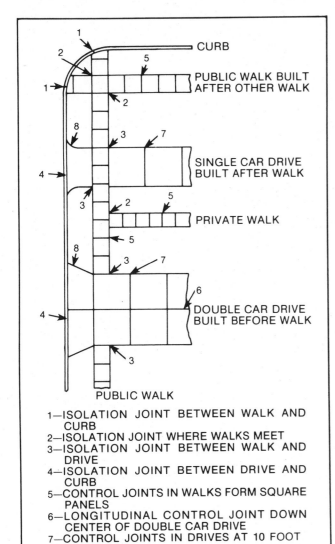

1—ISOLATION JOINT BETWEEN WALK AND CURB
2—ISOLATION JOINT WHERE WALKS MEET
3—ISOLATION JOINT BETWEEN WALK AND DRIVE
4—ISOLATION JOINT BETWEEN DRIVE AND CURB
5—CONTROL JOINTS IN WALKS FORM SQUARE PANELS
6—LONGITUDINAL CONTROL JOINT DOWN CENTER OF DOUBLE CAR DRIVE
7—CONTROL JOINTS IN DRIVES AT 10 FOOT INTERVALS
8—RADIUS OR FLARE AT DRIVE ENTRANCES

Recommended locations of isolation and control joints in walks and drives.

Isolation joint between sidewalk and curb. Nails anchor the joint material to the slab.

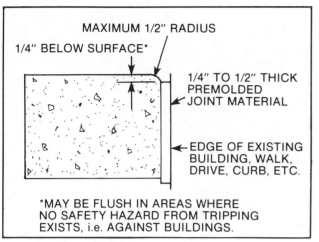

MAXIMUM 1/2" RADIUS

1/4" BELOW SURFACE*

1/4" TO 1/2" THICK PREMOLDED JOINT MATERIAL

EDGE OF EXISTING BUILDING, WALK, DRIVE, CURB, ETC.

*MAY BE FLUSH IN AREAS WHERE NO SAFETY HAZARD FROM TRIPPING EXISTS, i.e. AGAINST BUILDINGS.

Details of an isolation joint.

Stone and Brick Patios

A concrete slab is the best base for a surface of slate, flagstone, cut stone, or brick. Construction is simple and is basically the same for all the materials mentioned.

Build the slab as previously described, taking into account the extra height of the surface material. The masonry surface can be added when the concrete has cured.

Slate is available in random shapes and sizes or squared-off blocks. The latter is more expensive. Fit all the pieces together on an adjacent plot of the same dimensions as the original. Spacing will vary considerably when slates have random shapes and sizes. Maintain a minimum of a ½-inch space between adjoining stones. Flagstone is laid out in the same manner as random slates. Cut stone is set in a regular squared slate manner.

Before any of these masonry units are placed, thoroughly wet the surface of the concrete slab leav-

Concrete is a popular patio flooring material.

ing no puddles. Mix a batch of cement in the proportion of 1 part portland cement to 2½ parts sand, with enough water to make a firm consistency. Spread this with a trowel to a depth of about 1 inch, placing the slates about ½ inch apart, pressing them firmly into the mix. Cure slowly, sprinkling frequently with water. When completely set, mix a batch of grout with a more liquid consistency. Thoroughly moisten the crevices to be grouted, and work the grout into the spaces and level it with the surface. When it starts to set, finish the joints.

To avoid mortar stains, brush the slates with raw linseed oil before grouting. After the grout has set, spray the patio with water twice daily for two or three days before using. Mortar or concrete stains can easily be washed off.

Cut stone, flagstone, and brick can be laid in the same manner, except that bricks should be set with narrower mortar joints. The same procedure can be used for concrete slab sidewalks.

Stone Walks Flagstone and slate are frequently used for secondary walks, particularly in lawn and garden areas. The stones can be practically any shape, approximately 2 inches thick with their edges trimmed so that both sides of the walk are fairly uniform. Most approach walks have widths of 3 feet or less. Stones can be laid in a sand base,

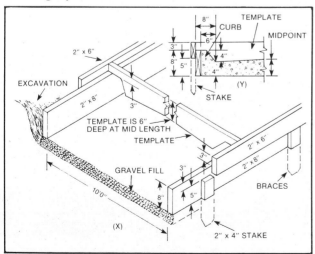

Typical forms for a convex driveway with curbs.

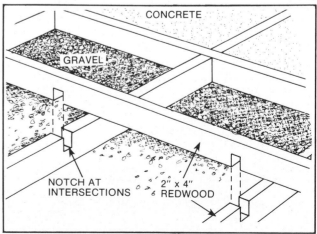

Notching for a permanent divider.

Slate and brick are popular masonry surface materials for patios.

but a concrete base with mortar joints is stronger and more durable. Cinder or gravel fills are suitable for clay or wet soils.

Brick Walks Brick sidewalks, that enhance landscaping, also make pleasing secondary walks. A sand subgrade gives a little as a person walks, thus providing more comfortable walking. When using the sand cushion, the bricks should have a decided pitch for drainage.

Concrete Precast Slabs

Concrete precast slabs, available from a variety of dealers, make ideal units for building walks. You can easily build a form and make your own slabs. In either case, the design, pattern, arrangement, and colors are unlimited.

To build the form for making the precast slabs, use ¾-inch exterior grade plywood as the base, cut to appropriate dimensions. Then, saw 1-inch boards into 1¾-inch widths for use as edge strips or lips around the four sides of the base. Saw more 1-inch boards into several beveled strips, ⅝-inch wide at the base and ⅜-inch wide at the top. These create the pattern in the form. Outline any desired ar-

rangement on the base as long as all the sides are straight. Nail the edge strips to the base; cut the pattern strips to the various lengths. Glue and nail them to the outlines you have made. Use wood filler to round off corners or joints where pattern strips meet and use it along the strips next to the base. Sand the edges and surfaces when dry. Finally, moisture-proof the form with three coats of marine spar or synthetic varnish. For added protection against moisture and easy removal of the slabs, brush oil on the form before each casting operation.

Place the form on a level surface. Prepare a mix of 1 part portland cement, 2 parts sand, 3 parts fine gravel, and the recommended amount of coloring powder, adding water until you achieve a workable consistency. Shovel or pour about ½ inch of the mixture into the oiled form. Place wire mesh cut to size in each flag mold section. Add more cement, and strike it level with the tops of the pattern strips and edge strips. To raise the level slightly, sprinkle gravel into the mix. Tamp or compact the mixture, eliminating air bubbles and voids, by raising each corner of the form about 4 inches and dropping it. Do not trowel the mixture. The pebbly surface will

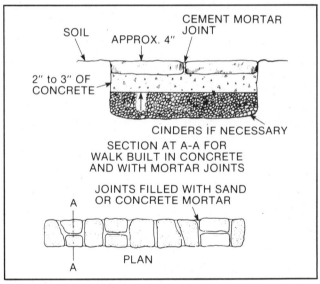

Details of stone walk.

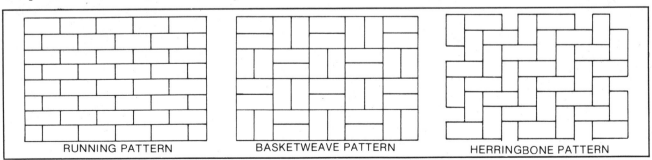

Popular brick patterns.

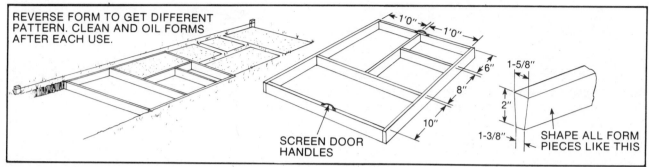

Reusable forms for making concrete blocks or flagstones.

assure a better bond when installed. Wait 48 hours for partial curing and shrinkage before removing the stones from the form.

To simplify removal, place an equivalent-sized sheet of ¾-inch plywood on top of the form. Turn it gently onto its face and insert thin wood strips under the four corners. Tap the form's back with a hammer to dislodge the flagstones. Carefully raise the emptied form.

Gently transfer the stones to a location where they can cure thoroughly. Cover them with moist burlap, canvas, or plastic sheeting for about four days.

The slabs can be colored. The illustrations show how the slabs are made and laid on a concrete surface.

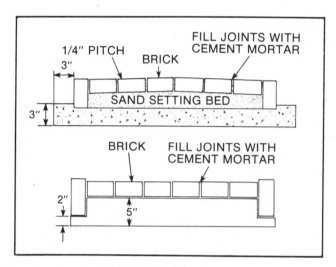

Two methods of laying brick walks.

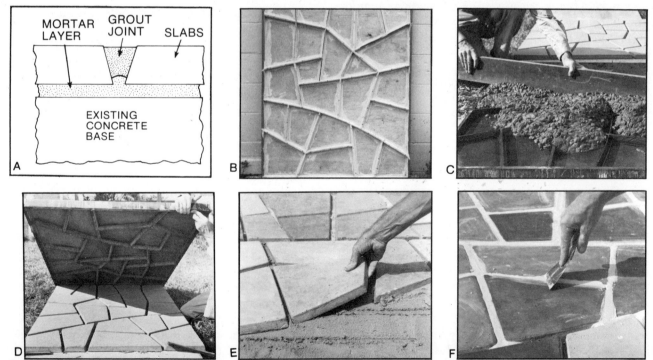

Steps in making concrete slabs and laying them on an existing concrete base. After the concrete is poured into the form (B), it is leveled with a straightedge (C). After the concrete has cured, the slabs are removed from the form (D). The slabs are set on existing concrete surfaces as shown in Steps A, E, and F.

Masonry Projects for the Home

Concrete and masonry products can be used for many projects about the home. Most of them utilize the masonry techniques described in this book.

Garden Walls

Garden walls and fences of masonry units can take on many delightful forms to enhance the landscape. They can be built with solid or screen block, bricks, or stones.

The choice of material depends primarily on the effect you wish to create. For example, the soft, warm tones of brick make a perfect foil for the varied greens of plants and vines that can be grown near the base. Either common or face brick may be used. Brick walls are usually made of the cheaper grades of hard, common brick. Soft bricks have a tendency to crumble.

Fieldstone or even cut stone can be used to make a garden wall. The individual stones, particularly rubble or fieldstone, should be selected carefully. Each stone in the wall should remain firmly in place, even without mortar. Keep round or ovoid stones to a minimum. Every stone should be placed to support the adjacent stones.

Concrete block garden walls are easiest to make, and with specially designed screen wall units or grille block, concrete masonry offers a new dimension in screen wall design. These units come in many sizes, shapes, colors, patterns, and textures. Some designs are available only in certain localities, and others are restricted by patent or copyright. Look for units early in the planning stage.

Garden walls range from 3 to 8 feet in height.

Steps and Porches

When slopes to the house or garden area are greater than a 5 percent grade, steps are necessary. In many areas steps at entranceways must conform to the provisions of local building codes. These codes specify critical dimensions, such as: (1) width; (2) height of flights without landings; (3) size of landings; (4) size of risers and treads; and (5) relationship between riser and tread size.

Steps for private homes are usually 48 inches wide and should be at least as wide as the door and walk they serve. A landing is desirable to divide flights of more than 5 feet, and it should be no shorter than 3 feet in the direction of travel. The top landing should be no more than 7½ inches below the door

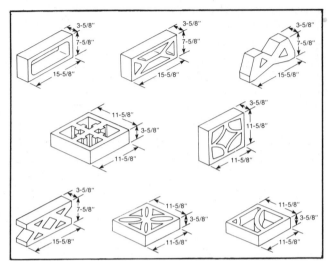

Typical screen wall units.

threshold. For flights less then 30 inches high, maximum step rise is usually 7½ inches, and minimum tread width, 11 inches. For higher flights, step rise may be limited to 6 inches, with a minimum tread width of 12 inches. Riser and tread size depends on step use. For aesthetic reasons, steps with risers as low as 4 inches and treads as wide as 19 inches are often built.

For optimum comfort and safety, the sum of riser and tread should equal 17½ inches. More generous steps may be desirable in leisure areas, such as patios, gardens, and terraces. In these instances, the following combinations of riser to tread dimension (in inches) can be used: 4 to 19, 4½ to 18, 5 to 17, 5½ to 16, or 6 to 15.

The closer the climbing step comes to the normal walking stride, the safer and easier it is for all ages. Treads and risers should be uniform in any one flight of steps.

Concrete Steps Concrete porches and steps are attractive, safe, durable, and not slippery in wet weather. The first step in making a small stoop and stair arrangement is the excavation. This should be 2 feet deep and roughly the size of the area to be covered by the steps. Before pouring, the excavation should be free of dirt and stones. The 2-foot earth sides will retain the concrete, so no forms are necessary below grade. The foundation of the building forms one end of the hole and must be thoroughly cleaned of all dirt. Tamp the earth in the hole to produce a sound base for the concrete.

When the excavation is completed, build the form for the steps. Side forms are usually 1-inch boards backed with 2 x 4 form studs, braced and tied. Riser forms for steps not more than 3 feet wide may be 1 x 8-inch boards. Wider steps require 2 x 8-inch riser forms to prevent bulging.

If 1 x 8 or 2 x 8-inch riser forms are used, an actual step height, or rise, of about 7½ inches is obtained, resulting in steps that are easy to climb. To make steps for maximum climbing comfort, the riser form boards may be tilted in at the bottom about 1 inch to provide additional toe space on the treads. Step edges may be rounded to help prevent chipping. Pour the foundation first and let it set for at least one day before doing the work above grade. Concrete for the foundation should be mixed with 1:2 ¾:4 formula.

After this foundation has cured for the required time, set the form on it. Level and square with the house by measuring the diagonals across the square that will form the steps. If both are equal, the sides are square, parallel, and ready to be fastened into position. Brace firmly into place with 2 x 4 stakes, and two to four diagonal braces running from the top edge of the sides to stakes in the ground wherever the form needs support.

When the form has been placed and secured, spread a thick cement paste over the base and house foundation to bond the new concrete to the old. Mix ½ bag of cement with 3 gallons of water and spread the paste evenly over the top foundation immediately before pouring the final batch of concrete, mixed with a 1:2¼:3 ratio. Pour carefully to avoid jarring the form out of position.

After the concrete has set for a few hours, finish the platform and steps with a wood float to produce

In these concrete and stone steps, low-profile risers, wide treads, and intermediate landings are combined to complement the sloping landscape.

a surface that will not be slippery in wet weather. Round off all exposed edges with a trowel.

Remove the form after a week, and cure the concrete by covering it with damp straw, burlap bags, or plastic for a week or 10 days. After this curing, the steps are ready for use.

Before building a form for patio steps, prepare the subgrade. If the soil is well drained and contains gravel and sand in relatively small amounts, just level and tamp the area thoroughly. If, however, the soil is a heavy, tight clay which provides poor drainage of surface water, excavate 6 to 10 inches below the grade and put in a thoroughly tamped gravel fill. Never use cinders as fill because the heavy concrete casting will crush them. After the forms are set, the concrete is mixed and poured in the same manner as described for porch steps.

Where the stairwell is already formed with concrete walls, such as for basement steps, riser forms

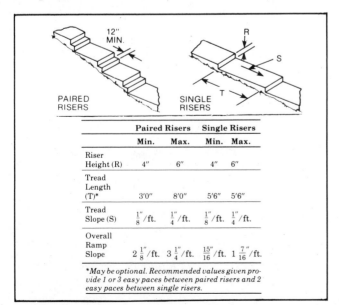

	Paired Risers		Single Risers	
	Min.	Max.	Min.	Max.
Riser Height (R)	4″	6″	4″	6″
Tread Length (T)*	3′0″	8′0″	5′6″	5′6″
Tread Slope (S)	$\frac{1}{8}$″/ft.	$\frac{1}{4}$″/ft.	$\frac{1}{8}$″/ft.	$\frac{1}{4}$″/ft.
Overall Ramp Slope	2 $\frac{1}{8}$″/ft.	3 $\frac{1}{4}$″/ft.	$\frac{15}{16}$″/ft.	1 $\frac{7}{16}$″/ft.

May be optional. Recommended values given provide 1 or 3 easy paces between paired risers and 2 easy paces between single risers.

Details for stepped ramps.

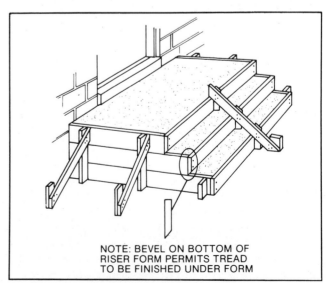

NOTE: BEVEL ON BOTTOM OF RISER FORM PERMITS TREAD TO BE FINISHED UNDER FORM

Typical concrete step form arrangement.

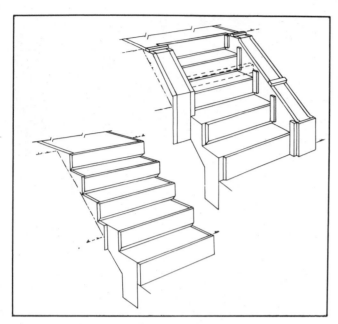

Forms for earth-supported steps. Parts of forms are cut away to show construction.

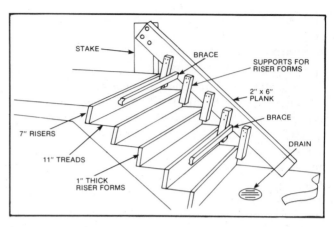

This type of form is used when building steps between existing walls. Notice the slanting risers.

spaced, plumbed, and wedged between the walls are the only forms needed. Earth fill is tamped to a uniform slope. When the steps are wider than 48 inches, it may be necessary to fit side planks and vertical cleats so that the riser boards are supported by horizontal braces wedged against the side planks. Keep the tread-to-riser proportions within the accepted comfort range. After the forms are set, 1:2¼:3 concrete mix is poured into them and finished as previously described.

Precast Concrete Steps   Precast concrete tread and riser units may be available at your local masonry supply dealer. They are easy to install, durable, and require no maintenance.

Precast concrete steps are supported by concrete-block walls placed on concrete footings, 6 inches thick and 12 inches wide, poured on firm soil below the frost line. Mortar is used in setting precast support walls, tread and riser units, platform units, precast stair stringers, and block support walls. The mortar should be made in the proportions of 1 part masonry cement and between 2 and 3 parts mortar sand. The sand should be damp and loose. Mortar joints should be thin and pointed up by tooling after the mortar has partially set. Mortar droppings should be wiped clean immediately with a wet cloth. Cure in the usual manner.

Brick Steps   Making brick steps is more difficult than constructing concrete units. Steps are generally laid in a bed of mortar or concrete. Excavate 9 inches and put in a foundation of gravel. Tamp down well. A concrete base is necessary so a

form must be constructed. A workable formula to use for brick steps is: twice the height of the riser plus the width of the tread equals 25. Steps should be at least 3 feet wide.

Pour a light concrete mixture of 1 part cement, 2¾ parts sand, and 4 parts pebbles into the form. The concrete base can be eliminated, and the solid brick bed can be used as a form if the steps are attached to the foundation walls with steel reinforcing. The excavation should be deep enough to accommodate a base of gravel and the thickness of two bricks laid flat. Sprinkle the bricks with a hose for five minutes before using.

Starting at one corner, place a bed of mortar on the foundation with a trowel, edge each brick into position, and tap down. Mortar must completely fill the space between the brick and the foundation. Cut off excess mortar between the joints. The second layer is laid in the same manner. Level and plumb frequently.

Finish the joints on the exterior face after the mortar has stiffened slightly. The weather joint is preferred because of its water-shedding ability.

The inside cavity between the side brick wall, the steps, and main body of the house is filled in with a mix consisting of 1 part cement, 2 parts sand, and 4 parts gravel. Steel reinforcing rods, ½ inch in diameter, give greater support.

Garden Steps   For garden steps, flagstones can be overlapped to make a very attractive series of steps. The ground should be removed carefully to permit the flagstones to be set solidly into place. Position them with a slight downward slope for proper drainage.

Concrete Floors   If a new concrete porch floor is to be no more than 2 feet above ground level, a simple slab built on a fill makes an excellent floor. The fill of gravel or crushed rock which supports the

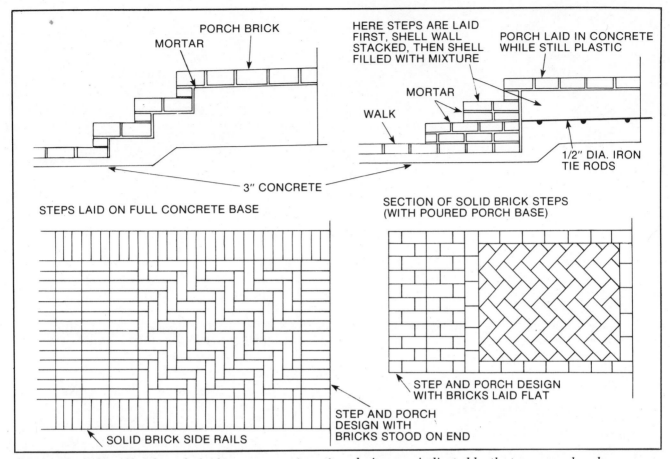

Steps and porches of brick can be laid in a great variety of designs, as indicated by the two examples above.

floor should be well tamped. The concrete floor is usually 5 to 6 inches thick and should be reinforced to prevent cracking. The thickened edges of the concrete slab should be reinforced with two ½-inch round reinforcing bars which creates a reinforced concrete beam that spans the distance between supporting piers. The piers are made by filling 8-inch post holes with concrete. At laps, the ends of ½-inch round bars should extend past each other about 2 feet. The porch floor should be sloped about ¼ inch per 1 foot to provide adequate drainage. The surface of a porch may be finished in the same manner as sidewalks and driveways.

Stone Planters

There is no limit to the size or shape of stone planters, and it requires no special techniques to get pleasing results. Any type of natural stone can be used.

The first step in building a stone planter is to excavate an area alongside the house foundation about 3 or 4 inches deep and 20 to 24 inches wide. Waterproof the foundation with an asphalt compound.

Fill this excavation with concrete, reinforced with two lengths of ⅜-inch steel rod near the base. Use a 1:2:4 mix for the concrete, and allow it to cure a day or more before resuming work.

Stones should be flat with straight faces. You can split larger stones and dress them with a hammer and chisel. Fit and position the stones on the slab providing for a minimum amount of joint space.

Thickness of Slab and Diameter and Spacing of Rod Reinforcement Required for Supported Concrete Porches*			
		Reinforcing rods	
Width of porch (feet)	Slab thickness (inches)	Diameter (inches)	Spacing (inches)
4	5	⅜	7½
6	5	⅜	6
8	5½	½	9½
10	6	½	8

*The transverse reinforcing rods are ⅜-inch round rods spaced 8 inches on center.

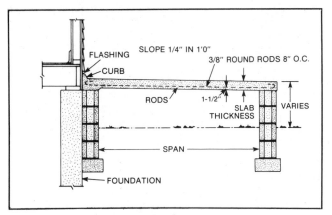

Construction of concrete porch.

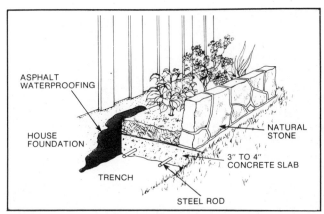

Details of a stone planter.

Spread mortar (1 part mortar cement, 2½ parts sand) over the outside edge of the slab and set the bottom row of stones into it. Set additional stones above this row wherever necessary to make the top edge of the planter straight. After all of the stones are in place, clean the joints with a bristle brush and fill them with mortar. Allow the mortar to cure for a day before filling the planter with soil and plantings.

Retaining Walls

Poured concrete, concrete blocks, brick, and cut stone can be used to build a retaining wall to prevent soil of sharply sloping lawns from eroding. Masonry retaining walls are most effective, practical, and permanent.

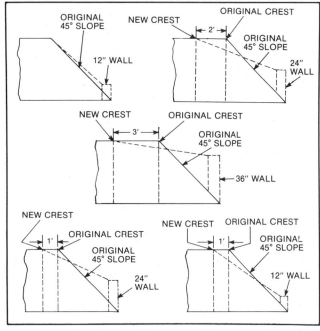

Various results obtained by using different wall heights and lawn crest positions.

Concrete Retaining Walls These are simple to build and the cost is moderate when compared with the benefits provided. First, determine the height of the wall. To assure stability, the angle of a sloping lawn should never be greater than 45 degrees—the gentler the slope, the better. The higher the wall, the farther forward the lawn crest will be. Various results can be obtained by using different wall heights and lawn crest positions.

The following illustration shows a cross section of a typical concrete retaining wall, and the chart indicates vital dimensions. This design does not require steel reinforcement because the width of the base and the weight of the unit provide adequate support.

It is important to provide adequate drainage for the soil behind a retaining wall to avoid excess pressure. Good drainage is accomplished by filling coarse gravel behind the wall and by building weep holes into the wall. Weep holes are made by inserting short lengths of 2-inch steel pipe or 3-inch drain tile in the forms when they are built. In walls built of stone or masonry units, the drain pipes can be put in place as the wall is built. For proper spacing, place the first row so that its outlets are 2 to 4 inches

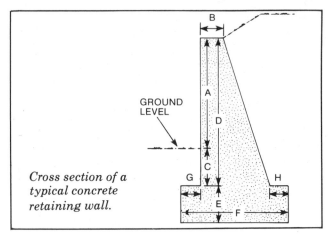

Cross section of a typical concrete retaining wall.

Retaining Wall Construction Data							
Exposed Wall Height (A)	Top Thickness (B)	Distance From Ground To Base (C)	Distance From Top To Base (D)	Base Depth (E)	Base Width (F)	Outside Base Extension (G)	Inside Base Extension (H)
12″	6″	4″	16″	14″	6″	3″	3″
18″	6″	6″	24″	18″	6″	3″	3″
24″	7″	8″	32″	24″	8″	4″	4″
30″	7″	10″	40″	28″	10″	4″	4″
36″	8″	12″	48″	36″	12″	6″	6″
42″	8″	14″	56″	40″	12″	6″	6″
48″	9″	16″	64″	44″	12″	6″	6″

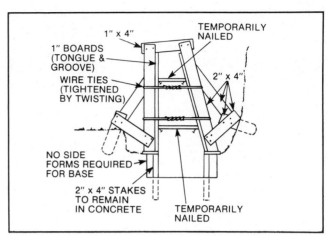

Details of a concrete retaining wall form.

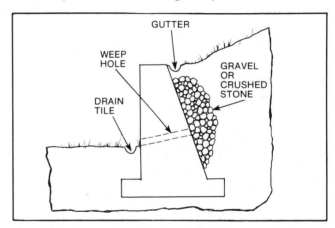

Completed concrete retaining wall with proper drainage.

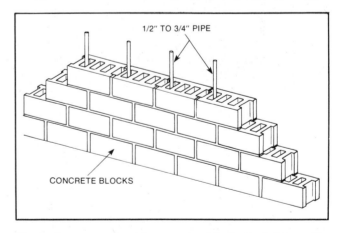

Typical construction of a concrete block retaining wall.

above ground level. If the wall is more than 4 feet high, place another row about 3 feet above the first row. Space holes 4 to 6 feet apart horizontally, and stagger vertical rows.

To produce a straight wall without bulges, forms must be rigid and well braced. The boards should be smooth on one side and have tongue-and-grooved edges. Duplex nails should be used in fastening the forms to ensure easy removal.

If the earth is sufficiently firm, the wall base may be cast in a straight-sided trench; otherwise, form boards will be required.

Both wood spacers and wire ties hold the form faces the correct distance apart. If the wall is not to be painted or covered with stucco, oil-coated form faces can be used to prevent the forms from sticking to the concrete.

In preparing concrete for retaining wall construction, use a 1:2¾:4 mix. Place the concrete in the forms in layers, 6 to 10 inches thick, in a continuous operation. Spade the concrete thoroughly next to the form faces to ensure smooth, even wall surfaces. Avoid excessive spading. If the concrete appears too wet and sloppy, add slightly more sand and gravel in succeeding batches. Do not vary the quantity of water. As pouring progresses, remove the wood spacers to prevent them from being covered by the concrete.

When the forms are filled, level the concrete. Finish after it has stiffened somewhat. If water appears on the surface of the concrete, allow it to evaporate before the final troweling.

Do not remove the forms for at least a week. Keep the concrete and the forms constantly moist.

The back of the wall should be waterproofed with an asphalt waterproofing compound. Tamp crushed stone fill into the space behind the wall. The top foot or so can be filled with topsoil. Provide a gutterlike depression in the soil along the wall.

Other Masonry Retaining Walls Concrete building blocks, if not laid too high, make good retaining walls. The cores are aligned and filled with cement or heavy pipe to key the blocks together. A wall of brick on a solid concrete footing can be used if the wall is not more than 4 feet high. Stagger the bricks inward with each course, leaving weep holes every 5 feet.

Walls of cut stone laid in mortar, or fieldstone and concrete, are sturdier than brick walls because their width is usually greater than standard size brick units.

Only simple forms of used lumber are needed for fieldstone and concrete walls. Footings and back foundation walls are built of concrete mixed and placed in the usual manner. The recommended mix for footings and foundation is 1 part portland cement to 2¾ parts sand to 4 parts gravel or crushed stone. For this mix, use 5½ gallons of water for each sack of cement when using moist sand. The concrete in which the fieldstone is set and which is used to back up the fieldstone is mixed in the proportions of 1 part portland cement to 2½ parts gravel. Use 5 gallons of water per sack of cement. The mix should be just stiff enough to remain in place without flowing.

The footing must be poured first. Forms are used only for the inner face of the wall. The studs or upright pieces can be 2 x 4 or 2 x 6 pieces spaced about 3 feet apart around the building and running the full height of the wall. These uprights are carefully plumbed and strongly braced. Walls are built in courses, usually about 2 feet high. Two 1 x 12-inch guide boards are used. These are raised as the walls are built, but should not be shifted until the concrete hardens enough to stand without slumping. This usually takes about a day.

Barbecue Fireplaces

For pleasant evenings and outdoor cooking, inexpensive and serviceable fireplaces can be built with minimum labor and time.

The outdoor fireplace can be as simple or as elaborate as individual taste prescribes. It can be any size and made of concrete blocks or masonry, reinforced poured concrete, a combination of brick and concrete, or even concrete faced with cobblestones. It can be finished with stucco or paint. Cobblestones and brick require a 2- to 3-inch concrete backing.

Whenever possible, position the fireplace so that the opening will face the prevailing wind. This will provide good draft and keep the smoke out of the eyes of guests. The two most common types of fuel are wood and charcoal briquets. If wood is used, do not build the fireplace beneath a tree or close to shrubs, because the excessive smoke will damage foliage. Gas and electric-fired outdoor barbecue grills are also available.

A proper foundation or base is essential for an outdoor fireplace. Construct the slab base at least 2 inches wider, on all sides, than the fireplace. If the soil in your yard does not have good drainage or is subject to frost heave, place the slab on a 4-inch layer of gravel or crushed stone. Build the slab as previously described.

The simple brick fireplace shown in the illustration is lined with standard fire bricks for increased durability. The bricks measure 9 by 4½ by 2½ inches and are laid so that the wide face is exposed to the flame. Use air-setting, high-temperature cement mortar. Thirty-five pounds will be needed for 100 bricks.

A prebuilt fireplace unit, simplifies building and reduces construction time by half.

Fireplaces and Chimneys

Because the fireplace can be a dominant and interesting exterior architectural feature as well as a focal point of interior design, fireplaces and chimneys should be aesthetically pleasing as well as functional.

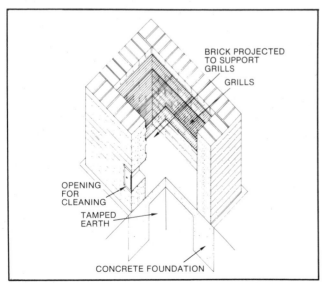

BRICK PROJECTED TO SUPPORT GRILLS

GRILLS

OPENING FOR CLEANING

TAMPED EARTH

CONCRETE FOUNDATION

Simple brick barbecue fireplace.

Various fireplace and chimney requirements are normally specified by local building codes. If the chimney is multistory, extra wide, or extra high, special design considerations are needed and such work should be left to the professional mason.

The design and construction of an efficient, functional fireplace requires consideration of fireplace location and the dimensions and placement of various component parts. The basic functions of a fireplace are: (1) to assure proper fuel combustion; (2) to deliver smoke and other products of combustion up the chimney; (3) to radiate the maximum amount of heat; and (4) to provide an attractive architectural feature.

Combustion and smoke delivery depend primarily on the shape and dimensions of the combustion chamber, the proper location of the fireplace throat and the smoke shelf, and the ratio of the flue area to the area of the fireplace opening. Heat radiation depends on the dimensions of the combustion chamber. Fire safety depends not only on design but also on the ability of the masonry units to withstand high temperatures.

Types of Fireplaces There are several types of fireplaces, but basic design and construction principles are the same. These types and standard sizes are given in the following table.

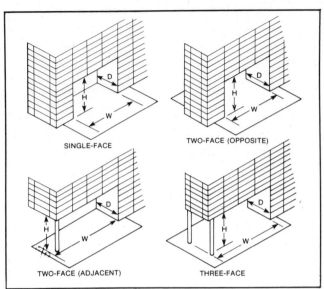

Types of fireplaces.

Fireplace Types and Standard Sizes					
Type	Width (w), in.	Height (h), in.	Depth (d), in.	Area of fireplace opening, sq. in.	Nominal flue sizes (based on 1/10 area of fireplace opening),* in.
Single-face	36	26	20	936	12 × 16
	40	28	22	1,120	12 × 16
	48	32	25	1,536	16 × 16
	60	32	25	1,920	16 × 20
Two-face (adjacent)	39	27	23	1,223	12 × 16
	46	27	23	1,388	16 × 16
	52	30	27	1,884	16 × 20
	64	30	27	2,085	16 × 20
Two-face (opposite)	32	21	30	1,344	16 × 16
	35	21	30	1,470	16 × 16
	42	21	30	1,764	16 × 20
	48	21	34	2,016	16 × 20
Three-face	39	21	30	1,638	16 × 16
	46	21	30	1,932	16 × 20
	52	21	34	2,184	20 × 20

*A requirement of the U.S. Federal Housing Administration if the chimney is 15 feet high or over; 1/8 ratio is used if chimney height is less than 15 feet. See Table 7-4 for nominal and actual flue sizes and inside area of the fireplace opening.

The traditional single-face fireplace is the oldest and most common variety, and most standard design information is based on this type. The multiface fireplace, used properly, is highly effective and attractive but may cause draft problems.

Fireplace Elements Only the major elements of a fireplace are discussed here.

- Hearth. The floor of the fireplace is called the hearth. The inner part of the hearth is lined with firebrick, and the outer hearth consists of noncombustible material such as firebrick, concrete brick, concrete block, or just concrete. The outer hearth is supported on concrete.

- Lintel. The lintel is the horizontal member that supports the front face or breast of the fireplace above the opening. It may be made of reinforced masonry or a steel angle.

- Firebox. The combustion chamber where the fire occurs is called the firebox. Its sidewalls are slanted slightly to radiate heat into the room, and its rear wall is curved or inclined to provide an upward draft to the throat.

  Unless the firebox is the preformed metal type, it should be lined with firebrick that is at least 2 inches thick and laid with thin joints of fireclay (refractory) mortar. Back and sidewalls, including lining, should be at least 8 inches thick to support the chimney weight.

  The fireplace is laid out on a concrete slab and its back is constructed approximately 5 feet high before the firebox is constructed and backfilled with tempered mortar and brick scraps. Do not backfill solidly behind the firebox wall to allow for some expansion of the firebox.

- Throat. The throat of a fireplace is the slot-like opening directly above the firebox through which flames, smoke, and other combustion products pass into the smoke chamber. The throat must be carefully designed to permit proper draft. It should not be less than 6 inches, and preferably 8 inches, above the highest point of the fireplace opening.

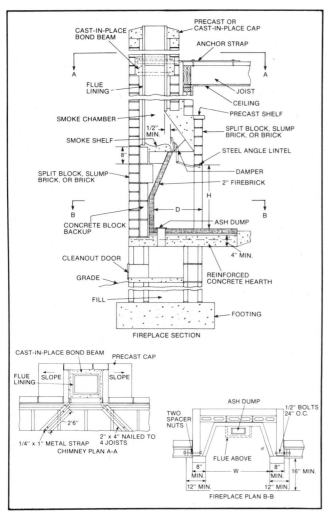

Details of masonry fireplace and chimney.

The inclined back of the firebox extends to the same height as the throat and supports the hinge of a metal damper. The damper extends the full width of the fireplace opening and should open upward and backward.

- Smoke Chamber. The smoke chamber compresses and funnels smoke and gases from the fire into the chimney flue. The shape of this chamber should be symmetrical with the center line of the firebox to assure even burning. The back of the chamber is usually vertical, and its walls are inclined upward to meet the bottom of the chimney flue lining. If the solid masonry wall is less than 8 inches thick, the smoke chamber should be parged with ¾ inch of fireclay mortar. Metal lining plates are available to give the chamber its proper form, provide smooth surfaces, and simplify construction.

Chimney Elements A fireplace or wood-burning stove chimney creates a draft and disposes of combustion products. Chimney design and con-

struction must assure efficient operation and freedom from fire hazards.

To prevent downward air currents, build the chimney at least 3 feet above a flat roof, 2 feet above the ridge of a pitched roof, or 2 feet above any part of the roof within a 10-foot radius of the chimney. Increasing the chimney height will improve the draft.

- Foundation. The concrete foundation for a chimney is designed to support the weight of the chimney. Because of the large additional weight, the unit bearing pressure beneath the chimney foundation should be approximately equal to that beneath the house foundation to minimize the possibility of differential settlement.

  The footing thickness should not be less than one and one-half times the footing projection. The bottom of the footing should extend below the frost line.

- Chimney Flue. A flue must have the correct area and shape to produce a proper draft. Smoke should be drawn up the flue at a relatively high velocity. Velocity is affected by the flue area, firebox opening, and chimney height.

  The total area of the flue opening should be approximately one-tenth of the fireplace opening area. Check your local codes.

A fireplace or stove chimney may have multiple flues, but each flue must be built as a separate unit. The American Insurance Association requires clay flue liners for all residential fireplace chimneys.

Flue linings extend from the top of the throat to at least 2 inches above the chimney cap. Chimney walls are constructed around the flue lining segments. The space between all masonry joints should be completely filled with mortar.

Modular-size chimney units, whether solid block or one-piece chimney units, can be combined only with modular-size flue lining. Minimum wall thickness measured from the outside of the flue lining should be 4 inches. Exposed joints inside the flue should be struck smooth.

When a chimney contains more than two flues, they should be separated by 4-inch thick masonry bonded into the chimney wall. The tops of the flues should have a height difference of 2 to 12 inches to prevent smoke from pouring from one flue into another.

Chimneys should be built as vertical as possible, but a slope is allowed if the full area of the flue is maintained throughout its length. Slope should not exceed 7 inches to the foot or 30 degrees. Where offsets or bends are necessary, miter both ends of abutting flue liner sections equally to prevent reduction of the flue area.

Decorative and Special Finishes for Masonry Surfaces

Many decorative finishes can be built into concrete during construction. Color can be added to concrete by using white cement, pigments, and colored aggregates. Textured finishes range from a smooth polished appearance to the roughness of a gravel path. Geometric patterns can be scored or stamped into the surface to resemble stone, brick, or tile. Redwood divider strips are used to form panels of various sizes and shapes. Special techniques can make concrete slip-resistant or sparkling. The possibilities are unlimited.

Colored Concrete

Concrete can be colored during construction with mineral oxide pigments and dust-on colors. Finished masonry surfaces can be colored with paints, dyes or stains.

Mineral Oxide Pigments   A wide range of colors is possible by adding mineral oxide pigments to the concrete finish mix. A single uniform color is most widely used in floors and walks.

Only pigments resistant to alkali should be used. Mortar colors containing a large percentage of filler are not suitable. Pure mineral oxide pigments and factory-prepared mixtures of cement and mineral pigment are available. Carefully follow manufacturer's directions. Mixing should be very thorough to achieve uniform color.

Added coloring materials should not exceed 10 percent by weight of the cement. Different shades can be obtained by varying the amount of coloring material used or by combining two or more pigments. Full color value can be obtained only with white portland cement. When pure white is desired, white sand and white cement should be used. White portland cement mixed with yellow and brown sands produces shades of cream, yellow, and buff.

Always mix color pigments and cement in the dry state. When using a mechanical mixer, mix all the ingredients for at least two or three minutes.

Best results are obtained by experimenting. After selecting the primary color, the exact shade can be determined by mixing a small sample of cement, pigment, and aggregate in the approximate proportions with enough water to give the mixture the consistency of syrup. Pour it on a nonabsorbent, rigid base to form a ½-inch thick layer. Place in a slow oven until it is thoroughly dry. Let it cool and then break it. The color of the fractured section is the sample color. By varying the sample proportions, the color may be altered until the desired shade is obtained.

Colored concrete is poured, worked, finished, and cured in the same manner as conventional concrete.

Dusted-on or Dry-shaked Colors   Horizontal concrete surfaces can be tinted by a dusted-on color mixture. This type of coloring is recommended only where foot traffic is light. Coloring by this method involves sprinkling and working in a dusted-on mixture before the poured concrete hardens.

After striking off to proper level, a dusted-on mixture of about 1 part cement, 1 to 1½ parts sand, and the required amount of pigment is applied immediately and uniformly at a rate of 45 to 125 pounds per 100 square feet of floor area.

After spreading the dry mixture, it should be wood-floated and worked into the slab until the surface becomes wet. Floating should be resumed when surface moisture disappears. Finish with a steel trowel. After 24 hours, cover the floor with a non-staining building paper to help curing. Cure the concrete in the standard way.

Painted Finishes   Many painting products have been marketed for use on concrete and concrete masonry. The basic constituents and most important characteristics of concrete and masonry paints are as follows:

- Portland cement paints are suitable for exterior or interior wall application, but are not acceptable for floors. A concrete or masonry surface must be damp for application, and setting and curing require water, a favorable temperature, and sufficient time for hydration.

  Prepared portland cement paints usually provide the best results (uniform color, durability, etc.). For best results, the surface should cure 3 weeks before painting.

- Latex paints are water-thinned, and dry very rapidly by water evaporation, followed by coalescence of the resin particles.

  Careful surface preparation is required for latex paints because they do not adhere readily to chalked, dirty, or glossy surfaces. They are easy to apply, have little odor and are economical, durable, nonflammable, breathing paints that are not damaged by alkalies. They have

Mineral Pigments for Colored Concrete Floor Finishes			
Color Desired	**Commercial Name**	**Approx. Quantities Required, in Pounds, Per Sack of Concrete**	
		Light Shade	**Dark Shade**
Grays, blue-black, and black	Germantown lampblack	1/2	1
	Carbon black	1/2	1
	Black oxide of manganese	1	2
	Mineral black	1	2
Blue	Ultramarine blue	5	9
Brownish red to dull brick red	Red oxide of iron	5	9
Bright red to vermillion	Mineral turkey red	5	9
Red sandstone to purplish red	Indian red	5	9
Brown to reddish brown	Metallic brown (oxide)	5	9
Buff, colonial tint, and yellow	Yellow ochre	5	9
	Yellow oxide	2	4
Green	Chromium oxide	5	9
	Greenish blue ultramarine	6	9

excellent color retention. Acrylics are more expensive than the other latex types, but generally perform better. Latex paints are suitable for exterior walls, but not for floor surfaces.

- Oil-based paints contain binding, drying oils. They are similar to conventional house paints, but they are nonbreathing. These paints are durable under some exposures, but are not particularly resistant to abrasion, chemicals, or strong solvents and can be damaged by alkalies.

  Oil-based paints are often modified with alkyd resins to improve resistance to alkalies, reduce drying time, and improve performance.

- Rubber-based paints form a nonbreathing film with alkali and acid resistance. They are suitable for floors, exterior surfaces, and interior surfaces that are wet, humid, or subject to frequent washing. Rubber-based paints may be used as primers under less resistant paints.

The success or failure of any paint depends on surface preparation. In most cases, success is more assured if the masonry surface cures at least six months before painting. If the paint is not sensitive to either moisture or alkalies, a long aging period is unnecessary.

Masonry surfaces must be free of dirt, dust, grease, oil, and efflorescence for paint to adhere properly. Dirt and dust may be removed by air-blowing, brushing, scrubbing, or hosing. Grease and oil can be removed by applying a 10 percent solution of caustic soda, trisodium phosphate, or detergents formulated for concrete. Efflorescence is cleaned off by brushing or light sandblasting. After any treatment, flush surface thoroughly with clean water.

Primer-sealers can be used to fill voids in open or coarse-textured masonry surfaces. The fill coats usually contain white portland cement and fine siliceous sand. If acrylic latex or polyvinyl acetate latex is included in the mixture, moist curing is not required.

Paint must be thoroughly stirred immediately prior to application. Paint thinning and color tinting should only be done in accordance with the manufacturer's directions for durable results. The paints commonly used on concrete masonry are applied by brush, roller, or spray.

- Portland cement paints are applied to damp surfaces by brush, with bristles no more than 2 inches in length. Scrub the paint into the surface. Allow 12 hours between coats. A 48-hour moist curing period is necessary after the final coat if the paint is not modified with latex.

- Latex paints may be applied to dry or damp surfaces by roller or spray, but a long-fiber, tapered nylon brush 4 to 6 inches wide, soaked in water for two hours before use is preferred. Dampen the surface if it is moderately porous, or dry weather prevails. These paints dry in 30 to 60 minutes and require no moist curing.

- Oil-based and oil-alkyd paints should not be applied during damp or humid weather, or when the temperature is below 50 degrees Fahrenheit. Application is by brush, roller, or spray to a dry surface. Allow each coat to dry at least 24 hours, and preferably 48 hours.

- Rubber-based paints are applied by brush to dry or slightly damp surfaces. Two or three coats are necessary. The first coat is usually thinned in accordance with the manufacturer's recom-

Apply paint with a roller to a masonry surface.

Portland cement paint is applied at the joints first.

mendations. Allow 48 hours between coats. Caution: Most paints are flammable; therefore, adequate ventilation should be provided during application.

Clear Coatings A clear coating applied to masonry will render the surface water-repellent and protect it from soiling and surface attack by airborne pollutants. In areas where weather exposure is not severe and air pollution is low, a clear coating may not be necessary.

The coating should be water-clear, capable of being absorbed into the surface, long-lasting, and not subject to discoloration. Coatings based on a methyl methacrylate form of acrylic resin are effective. Apply one or two coats as directed by the manufacturer.

Dyes and Stains Dyes and stains provide similar results, but differ in makeup and application. They will not rub off under hard wear but may let surface imperfections show through. Dyes are available in dark shades of green, blue, red, brown, and gray.

Thoroughly clean the surface and etch with acid to remove the slick surface and open the pores of the concrete. This permits easy penetration of the stain or dye. Acid etching is not necessary when using a stain on masonry walls.

Applying a Dye Mix 1 gallon of muriatic acid and 2 gallons of water for each 150 to 200 square

feet of surface area. Splash the solution on the floor and scrub it with a long-handled brush. Watch for the acid fizzing which indicates that the solution is working properly. Two applications of acid are usually required. Muriatic acid is corrosive, so be sure to follow the safety precautions on the container.

After each acid cleaning, flush the floor with plenty of clean water and examine it carefully. Spots which still feel slick to the touch must be cleaned again, using a stronger solution (1 gallon of acid to 1 gallon of water). Allow the floor to dry thoroughly, at least 1 week.

Approximately 1 gallon of dye will cover 150 to 200 square feet. Pour it into a flat pan, keep it stirred well, and apply with a long-handled brush. Work the dye into the surface twice. This is called fixing.

When the floor has dried overnight, examine it in good light. Use a common abrasive to remove excess dye from any glossy spots. Allow the floor to dry for at least a week.

Dressing completes the job and usually comes with the dye. Standard dressing is scrubbed on with a long-handled brush. Allow to dry 1 hour and then polish it with the same brush. Clean and dry the brush before you polish with it.

The only maintenance required is occasional dry-mopping or sweeping. When the dressing begins to wear, simply put more dye on the worn spots and buff as before.

Self-polishing dressing is brushed on in a full, wet coat and is permitted to dry overnight. A gallon covers about 400 square feet.

Applying a Stain Several types of stains may be used on concrete masonry. They include the three listed below.

- Oil stains require aging the wall for several months before application, or pretreatment of the surface to inhibit reaction between alkalies and the oil in the stain. Many stains suitable for wood are suitable for concrete masonry.

A clear coating can waterproof a brick wall.

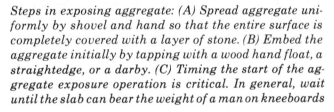

Steps in exposing aggregate: (A) Spread aggregate uniformly by shovel and hand so that the entire surface is completely covered with a layer of stone. (B) Embed the aggregate initially by tapping with a wood hand float, a straightedge, or a darby. (C) Timing the start of the aggregate exposure operation is critical. In general, wait until the slab can bear the weight of a man on kneeboards

with no indentation. Then, brush the slab lightly with a stiff nylon-bristle broom to remove excess mortar. (D) Next, fine-spray with water along with brushing. Special exposed-aggregate brooms with water jets are available. If aggregate is dislodged, delay the operation. Continue washing and brushing until flush water runs clear and there is no noticeable cement film left on the aggregate.

- Metallic salt stains are slightly acid solutions of salts that deposit colored metallic oxide or hydroxide in the surface pores. These deposits are not soluble in water.
- Organic dyes which contain analine dyes or indicators used in chemical analysis.

Staining should be delayed at least 30 to 45 days after the concrete masonry structure is built. The surface must be dry and clean. Two or more stain applications may be required to obtain the depth of color desired.

Each coat of stain should thoroughly saturate the surface and be evenly applied. Prevent the stain from overlapping a dried area. Allow 4 to 5 days between coats.

Plaster Finishes There are several acrylic compounds on the market that can be applied to concrete block masonry that will give a plaster- or stuccolike finish. Acrylic compounds are ready mixed, can be applied directly from the container, and are available in several colors. Preparation and application are much easier than with portland cement stucco and plaster.

Special Surface Treatments

There are several surface treatments including exposed aggregate, textured finishes, geometric patterns, sparkling finishes, and a combination of two or more surface treatments.

Exposed Aggregate Exposed aggregate is one of the most popular decorative finishes today. It offers unlimited color selection and a wide range of textures. Exposed-aggregate finishes are attractive, rugged, and highly immune to wear and weather.

The seeding method is one of the most practical and commonly used exposed aggregate techniques. See the accompanying illustrations for details.

In an alternate method, the aggregates are exposed in conventionally placed concrete. The mix should contain a high proportion of coarse to fine aggregate. The coarse aggregate should be uniform in size, bright in color, closely packed, and properly distributed. Concrete slump must be low. Usual procedures are followed in placing, striking off, bullfloating, or darbying. The aggregate can be exposed when the water sheen disappears, the surface can support a man's weight, and the aggregate is not overexposed or dislodged by washing and brushing.

Textured Finishes Interesting and functional decorative textures can be produced on a concrete slab with little effort and expense by using floats, trowels, and brooms. More elaborate textures are possible with special techniques.

A swirl finish lends visual interest as well as surer footing. The concrete is struck off, bull-floated, or darbied, and a hand float is worked flat on the surface in a semicircular or fanlike motion. Patterns are made by using a series of uniform arcs. Coarse textures are produced by wood floats, medium textures by aluminum, magnesium, or canvas resin floats, and fine textures with a steel trowel.

Broomed finishes are attractive, nonslip textures created by pulling damp brooms across freshly floated or troweled surfaces. Coarse textures are produced by stiff-bristle brooms on newly floated concrete. Medium to fine textures are obtained by using soft-bristle brooms on floated or steel-troweled surfaces. A broomed texture can be applied in straight lines, curved lines, or wavy lines. Driveways and sidewalks are usually broomed at right angles to the direction of traffic.

A travertine finish, or keystone, is created by applying a dash coat of mortar over freshly leveled concrete. The dash coat is mixed to the consistency of thick paint. Apply in a splotchy manner with a dash brush to form ridges and depressions. Allow the coating to harden slightly and flat-trowel to flatten the ridges and spread the mortar. The finish is smooth on the high areas and coarse grained in the depressed areas. Many interesting variations of this finish are possible, depending upon the amount of dash coat applied, the color used, and the amount of troweling.

Type of broomed finish.

A similar texture can be produced by scattering rock salt on the surface after hand floating or troweling. The salt is pressed into the surface so that only the tops of the grains are exposed. After the concrete has hardened, the surface is washed and brushed to dislodge and dissolve the salt grains, leaving pits or holes in the surface.

The rock salt and travertine finish should not be used in areas subject to freezing weather.

Geometric Patterns A wide variety of geometric designs can be stamped, scored, or sawed into a concrete slab to enhance the beauty of walks, drives, and patios. Random flagstone or ashlar patterns can be produced by embedding 1-inch strips of 15 pound roofing felt in the concrete in the pattern desired. The strips are carefully removed before the slab is cured.

Another method of producing these patterns is to use an 18-inch long piece of ½- or ¾-inch copper pipe bent into a flat S-shape. Score the surface with this tool while the concrete is still plastic, soon after darbying or bull-floating. A second scoring to smooth the joints is done after hand-floating.

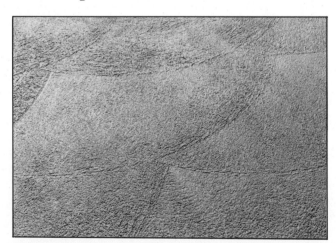

A swirl finish is produced with a hand float or trowel.

Typical travertine texture.

Random flagstone pattern on a sidewalk slab.

Divider strips and borders of wood, plastic, metal, or masonry create unusual patterns and designs. Allow concrete work to be completed in segments for better control of placing and finishing. Random cracks are greatly reduced or eliminated because the divider strips act as control joints. Wood divider strips should be made of rot-resistant lumber.

Plastic strips are difficult to install and must be securely staked.

Concrete masonry, brick, or stone divider strips and borders may be set in a sand bed, with or without mortared joints.

Nonslip and Sparkling Finishes Surfaces that are frequently wet or that could be dangerous if slippery can be given special nonslip finishes with floats, trowels, brooms, or dry-shake applications of abrasive grains.

Two widely used abrasive grains are silicon carbide and aluminum oxide. Silicon carbide grains are sparkling black and are also used to make sparkling concrete. The sparkle is especially effective under artificial light. Aluminum oxide may be gray, brown, or white, and is used where sparkle is not required or desired. Abrasive grains should be spread uniformly over the surface in a quantity from ¼ to ½ pound per square foot and lightly troweled, depending on the manufacturer's directions.

Combinations Striking effects can be created by combining colors, textures, patterns, and materials. Alternate areas of exposed aggregate can be combined with plain, colored, or textured concrete or other masonry units. Ribbons and borders of concrete masonry or brick add a distinctive touch when combined with exposed aggregate. Light-colored strips of exposed aggregate can divide areas of dark-colored concrete or vice versa. Scored and stamped designs are enhanced when combined with integral or dry-shake color. The options are limited only by imagination.

Strips of colorful brick add interest to this walk while acting as control joints to help prevent random cracking of the exposed aggregate surfaces.

Repairing Concrete and Masonry Surfaces

Concrete and masonry problems can usually be traced to the ingredients, proportions in the mix, or finishing techniques. If you carefully follow the procedures detailed in this book, concrete and masonry problems should be kept to a minimum. Troubles do occur, however, and they should be repaired immediately.

Repairing Cracks

Repairing Hairline Cracks Hairline cracks can be repaired with a grout made of portland cement and water. Add just enough water to form a thick paste.

Moisten the old concrete along the hairline crack with water for several hours before adding the grout to prevent the concrete from absorbing water from the grout and drying out the mixture. NO water should be standing on the surface when grout is applied.

Force the grout into the crack tightly with a putty knife or pointing trowel. Then, smooth it level with the original concrete.

After air drying, for about two hours, cover the patched area with a plastic sheet or a board for approximately five days. Lift the covering daily and sprinkle the repaired area with water.

Repairing Cracks in Slabs To patch broken, cracked, or crumbled portions of walks, driveways, patios, and floors, first chip away the bad sections and undercut the edges with a cold chisel and hammer. If the broken or cracked portions do not extend entirely through the concrete, roughen the bottom and undercut all sides. Undercutting helps to keep the new concrete bonded with the older concrete. For most concrete surfaces, undercutting should be done to a minimum depth of 1 inch. The depth depends on the size and depth of the crack to be repaired. After the crack has been thoroughly undercut, remove all loose material and brush the area with a wire brush. Use a garden hose to wash away the dust in the crack.

The strength of the patch can be increased if a concrete bonding adhesive is used. There are many types of concrete adhesives. Acrylic resin, a milky fluid, is one popular type. It should be brushed into the undercut area and allowed to dry until it becomes tacky. If you do not use a cement adhesive,

the area to be patched should be thoroughly brushed and soaked with water to prevent the old concrete from absorbing moisture.

Premixed concrete patch is suitable for small patching jobs. If you use ready-mix patch, you just add water. Gravel or crushed stone can be added for large holes. If you wish to mix your own mortar, use 1 part portland cement and 3 parts clean sharp sand mixed with enough water to make a rather stiff mix. For concrete patches of any size, a 1:2:3 mortar mix is recommended.

Fill the patch about ¼ inch above the adjoining surfaces to prevent the patch from becoming slightly concave after it sets. Allow patch to cure for about 2 hours before finishing. Move a steel trowel or float sideways with slight downward pressure. Raise the advancing edge a little. To hold concrete at the edge of a walk, use a flat board and stakes. When a patch crosses a joint, use a groover to continue the groove. Use an edger to round the outside edge next to the form.

Small, shallow holes in flat concrete surfaces require no additional chipping and can easily be repaired with latex cement. Clean the small hole with a wire brush to remove all of the loose concrete particles. After all concrete particles have been brushed away, thoroughly wash the area with a garden hose.

The latex cement usually comes in 5-pound cans. The liquid latex and cement are separated in the can. Pour the latex into the larger can and thoroughly mix with the cement to form a heavy paste. Apply this paste to the cleaned hole in approximately ¼-inch layers. Smooth out each layer with a trowel and allow to partially dry before applying

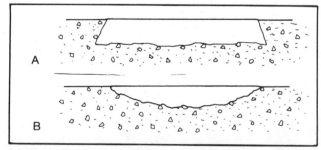

Method of making an undercut: (A) Right way—undercut the crack to give the necessary holding power to the new patching mix. (B) Wrong way—do not just pour new concrete into the old crack.

Concrete bonding adhesive can be applied directly to the concrete mix as directed by the manufacturer.

the next layer. Build up the latex cement until it is to the level of the original concrete. Finish as regular concrete.

For large holes or cracks, reinforcement of the concrete is necessary. Half-inch hardware cloth or 1-inch chicken wire may be used. Fill the hole halfway with concrete, tamp, drop in reinforcing wire, and tamp again. Then, fill the patch and finish.

After a few hours, when the surface has set hard enough to resist marring, lay a piece of tar paper over the patch and cover with sand or dirt to retard evaporation and permit proper curing for 2 days. Protect very large patches and holes for four or five days.

When a slab has sunk, the high edge can be chipped off if the difference is slight. A sunken slab can be lifted with a crowbar and relaid on a filling of crushed rock. Heavy driveway slabs may be cut back 12 inches or so, and the space between them can be filled with a concrete ramp.

If finished grooves are not scored deeply enough, the concrete cracks irregularly instead of beneath

Moisten the area to be repaired with concrete adhesive or water.

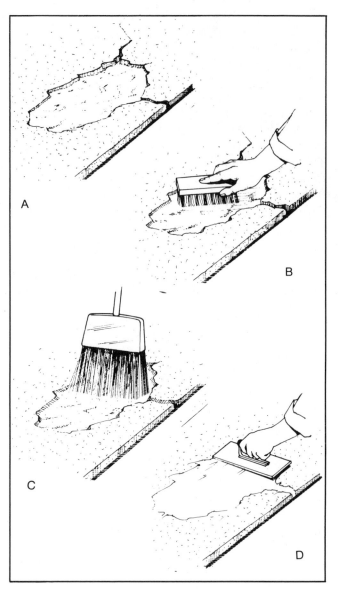

(A) Small holes in concrete surfaces can usually be repaired with latex cement. (B) All loose particles should be brushed away before latex cement is applied. (C) After brushing away loose particles, wash out with a garden hose. (D) Finish off the concrete with a trowel.

the groove. When the slabs are small enough to handle, use a hammer and stonecutter's chisel to deepen the groove. Gradually strike harder blows, especially near the ends, until the piece breaks clean. Level the two trimmed slabs into position and fill in with a square of new concrete.

When a sidewalk buckles, cut out one or more slabs at the highest point and relay them, adding an expansion joint. You can pour new slabs, molding an expansion joint in one.

When replacing one or more slabs of walk, break the old concrete with a heavy sledge. Wear safety glasses, as the chips will fly. After removing the debris, excavate to proper depth and erect forms to hold the concrete. You can use some of the old pieces of concrete to take up space but allow enough space between pieces to ensure proper bonding. Wet the rubble to prevent rapid absorption of moisture from the new concrete. Using a concrete bonding adhesive is advisable.

Where the new slab is subject to stress, connect it to the existing slab with steel reinforcing rods. These rods should be set into holes drilled about 1 foot apart into the edge of the existing slab. Place them 1 foot from the sides and 6 inches from the ends extending at least 6 inches into both slabs.

If the concrete requires extensive patching, resurfacing the entire area may be necessary. Proportions for mixing resurfacing concrete are 1:1:1½. The mix can be made more workable by changing the amount of sand. Avoid adding more than 5 gallons of water per bag of cement.

Erect the forms and stake them tightly against the existing concrete. Set the tops to the desired new slab height, about ½ inch to 2 inches above the old one. Before pouring, cover the area with a concrete bonding adhesive. Place and finish the concrete in the normal manner.

Chipped Concrete Steps   If the overhanging front edge of a concrete stair tread is chipped or broken, undercut the sides of the crack or break off the damaged areas. Force the patching concrete,

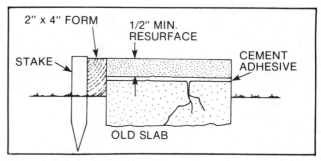

Form for resurfacing a walk or drive.

mixed to the proper consistency, into the enlarged cut. Set a form made of either wood or thin-gauge sheet metal under the overhanging edge to support the patch while it is settling and curing. The same technique can also be used to repair broken corners.

Repairing Cracks and Holes in Concrete Walls   Repairing structural cracks in a concrete wall requires basically the same steps as repairing a crack in a slab. Undercut and widen the crack with a cold chisel and hammer as previously described.

After all loose material has been chipped away, thoroughly clean the undercut area with a stiff wire brush. Do not smooth off the edges. The rough surface created by the chiseling provides a good bond.

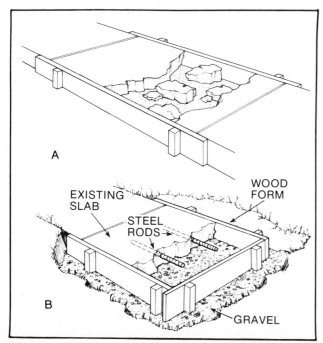

(A) Form used when replacing a broken slab of a concrete walk. (B) Method of reinforcing or tying a new slab to an existing one.

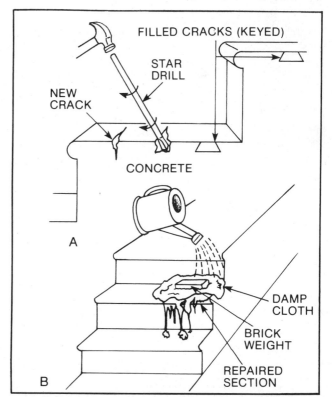

Repair procedures for various concrete step problems. (A) The crack must be undercut before filling. (B) To prevent recracking, the new concrete patch must be properly cured.

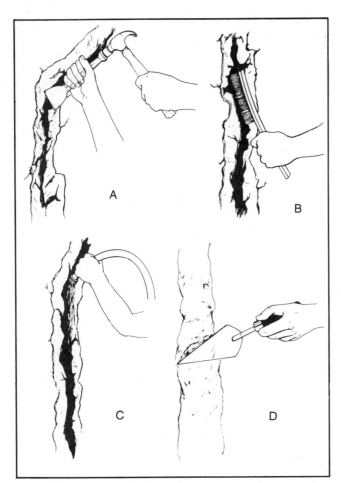

Repairing a structural wall crack: (A) Enlarge and undercut the crack with a cold chisel and hammer. (B) Clean out the undercut with a wire brush. (C) Moisten the area to be repaired to prevent the newly applied patch from drying out. (D) Apply the concrete to the patch with a pointing trowel.

Apply cement adhesive to the enlarged area after it has been thoroughly cleaned. If you do not have a cement adhesive, prime the area with a thin, creamy mixture of portland cement and water. Although moistening the area before filling may be adequate, concrete adhesive or the portland cement and water mixture is much more desirable.

Ready-mix concrete patch can be used for small cracks in cement walls. This mixture should be forced tightly into the crack with a pointing trowel. The finish on the old concrete is often difficult to duplicate. To conceal the patch, experiment by roughing up the patched area while it is still workable until the desired finish is created.

Holes and broken areas in concrete walls can be patched in the usual manner; undercut the area, thoroughly clean it, and apply cement adhesive, and force the patch mix into the hole. Moisten and cure the area after the patch is applied.

Dampproofing a Masonry Basement

The first problem in attempting to dry a wet basement is to determine the cause of the dampness. Condensation caused by the contact of warm, damp air with the cool masonry of the walls and floor is common. You can readily determine whether condensation is the cause by sticking a sheet of aluminum foil to the wall in the center of the wet area. If the outer surface of the foil remains dry after 12 hours, condensation is not the cause of dampness. If drops do appear on the outside of the foil, condensation is the cause. Leakage from a wall cannot pass through the foil.

A leak is generally easy to recognize as a crack in the wall, an open joint between walls and floor, or porous sections in mortar. The extent of leakage depends on the water pressure working against the crack. When a house is built on low and soggy ground, the pressure is great enough to force water through tiny cracks in the walls and floor, or through porous concrete.

Pressure will depend on the outside conditions and natural drainage. There will be little or no dampness in a cellar if the earth below it is sandy or loose and the house is on high ground because water is absorbed too rapidly to accumulate and to exert pressure. Leaks occur when the water forms pools against the foundation walls and works its way through the walls.

Surface drainage can result in water problems. If surface water is entering your basement after heavy rains, it generally reveals itself by stains on the wall starting at a level with the outside grade and gradually diminishing near the floor.

Never allow water from gutters to flow directly from the downspouts onto the earth. A concrete slab

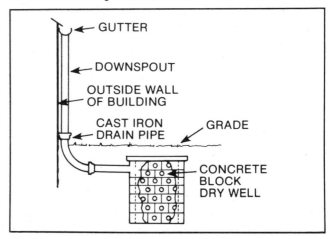

Typical dry well arrangement.

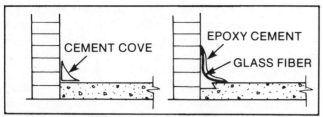

Methods of sealing the joint between the basement floor and the wall.

or splash block beneath the downspout only partially deflects water. It will not keep the water out of your basement. The only permanent solution is to build a dry well as shown in the illustration.

Quick drainage away from the house is essential. This can usually be accomplished by proper grading. Place additional filling against the basement wall and grade it to a sharp, smooth slope that extends at least 8 or 10 feet from the wall. If it is necessary to grade above basement windowsills, an areaway can be built around them. You can use asbestos sheets, masonry, or ready-made metal areaways. Line the bottom with cinders, gravel, or crushed stone to provide drainage. Plant the soil around the foundation generously to help resist erosion.

Another method to provide drainage is to lay a concrete pavement, walk, or gutter, 2 or 3 feet in width, around the house, with a gradual slope away from the walls. Where the sidewalk joins the wall,

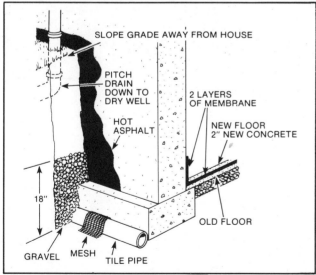

Several of the more popular methods of waterproofing a basement. One of the best is to lay tile pipe around the foundation footing. The walls should be coated with hot asphalt. All gutters should be connected to dry wells, and the grade should be sloped away from the house. Inside the house, a floor often must be laid over the old one, once an asphalt membrane has been applied as shown.

the wall surface should be roughened, cleaned, and moistened, and the concrete rounded to meet the face of the wall to create a good bond.

Interior Treatments If your basement is still damp after these precautions, work special waterproof coatings and waterproof cement into inside wall surfaces. These powdered waterproofing compounds are most effective in stopping seepage, as well as in checking leaks in masonry walls. Among the most popular varieties are the heavy-duty cement compounds, which are mixed with water before application. They are much heavier in consistency than powdered cement paint and have sealers added to prevent seepage. They can be applied only over an unpainted masonry wall or walls previously painted with ordinary powdered cement paint. Clean walls thoroughly before applying.

While cleaning the walls, check for holes or cracks. Any crack or hole must be filled before the waterproofing is applied. Cut the crack with a cold chisel to form a wedge, with the inside wider than the outside to prevent the patch from falling out after it dries. Use a stiff wire brush to remove any dirt or loose concrete from inside the crack. Wet the sides of the crack and mix a mortar using a quick-setting hydraulic cement. Force the mortar in to completely fill the opening, and smooth the outside surface. Keep the patch moist until it sets.

After the walls have been cleaned and all cracks filled, the waterproofing material should be applied as directed by the manufacturer. Powdered waterproof compounds require wetting the wall before application. One coat is usually sufficient to do the job. Two coats may be necessary where seepage is severe.

Other popular waterproofers are the "blockfillers." Some have an alkyd or a modified oil base, while others use special latex or plastic binders. Like the powdered types, these materials should be applied only over a properly prepared surface. Blockfillers are only suitable for mild seepage problems or occasional leaks.

Epoxy-type sealers are effective in hydrostatic pressure. Most hydrostatic leaks appear along the joint where the floor and walls meet. Apply epoxy sealer along this area. Brush on two coats, extending about 1 foot from the bottom and about 3 inches onto the floor. Apply the epoxy to porous sections or other problem areas.

If the basement floor has a dampness problem, waterproofing depends on its condition and use. If the floor leaks periodically through hairline cracks, it can be treated with blockfillers (if the floor is unpainted) or with an epoxy sealer (if the floor is painted or if the problem is hydrostatic in nature). If the basement floor is to be covered, the wet floor can be treated with asphalt and felt. Apply a base coat of either asphalt or coal-tar pitch to the floor and a layer of 15-pound roofing felt or polyethylene plastic sheeting over it. Before the bituminous coating is brushed or troweled on, prime the floor with an asphalt primer with asphalt coatings and a coal-tar or creosote primer with coal-tar coatings. When laying the felt, make certain that all bubbles and bulges are ironed out.

A badly cracked and leaking floor should be replaced. Top the existing floor with a 2-inch layer of 1:2:3 mix concrete reinforced by lightweight No. 14 gauge, steel-wire mesh. To prevent cracking, keep the floor moist for at least three days.

Immediately before pouring the new concrete, roughen, clean, and wet the floor. A cement-sand grout (1 part cement to 2 parts sand volume) should be scrubbed in to form a bond. If the old floor has been penetrated by seepage, a bituminous coating should be applied to the surface of the old floor before the new concrete topping is set in place.

Exterior Treatment Should interior basement repairs fail to stop dampness, the only recourse is exterior waterproofing. Dig a trench around the foundation as deep as the footings. A line of 4-inch diameter plastic or clay drain tile should be laid at the bottom of the trench at a pitch of 1 inch in 4 feet. The tile should be connected with a sewer, dry well, or other outlet at a lower level. Joints should be covered with strips of roofing paper or tar paper to prevent sediment from running into the pipe. The

pipe should be carefully laid and protected against settling or leakage by tamping fine-screened gravel or broken stone around it. Following this, coarser gravel, up to 1 inch in size, should cover the pipe to a depth of 1 or 2 feet. Before backfilling with earth to grade level, spread burlap or bagging, or place sod, grass side down, on top of the stone to prevent fine material from washing down into the stone.

Apply a layer of bituminous coating to the exposed foundation wall, even if this was done as part of the original construction. Before applying the coating, thoroughly clean the surface. Brush or trowel on a heavy layer of bituminous coating, covering this with sheets of roofing felt, overlapped at least 10 inches. To complete the waterproof membrane covering, apply another coat of asphalt or coal-tar pitch. Allow this to harden for at least a day before replacing the soil carefully.

Solving Problems in Brick Walls

A properly constructed brick wall should be waterproof under normal conditions. Leaks are most common in the joints between bricks. Vertical mor-

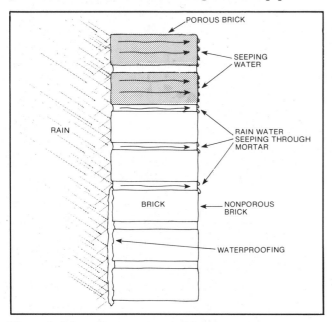

How a brick wall can cause problems.

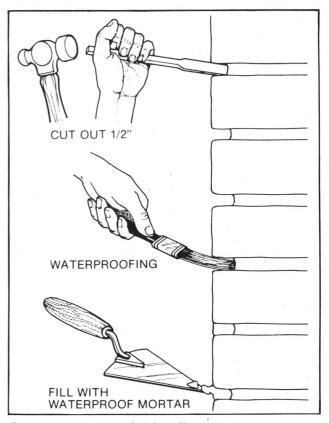

Steps in repointing a brick wall.

tar joints must be packed tightly, and the mortar must be brought out flush with the surface of the brick. On horizontal joints, the mortar is sloped to shed water. If there are any depressions in the joints, water can collect in them and penetrate the smallest crack in the wall. More cracks will occur when the moisture freezes and expands during the winter.

Where mortar joints have softened or disintegrated, cut out all loose or disintegrated mortar joints to a depth of at least ½ inch and repoint or fill with mortar. If the work is being done to correct leakage, all joints should be cut out in the affected area. If the repointing is being done as maintenance work, remove only the defective mortar.

Use a cold chisel or small pick to clean out the loose mortar, but do not damage the sound mortar or the bricks. After cutting out the mortar, brush out the cavity and wet it with a fine spray of water. Mix a cement mortar of 1 part cement to 3 parts fine sand and pack this into the joint. Trowel it smooth and build out the lower portion slightly.

If an entire brick becomes loose, remove all the mortar around the brick with a cold chisel, taking care not to damage other joints or bricks. Clean out the wall cavity and wet the sides. Soak the replacement brick in water for a short time. Spread cement mortar generously around the sides of the opening and over the brick. Force the brick back into place and pack the joints tightly with mortar; smooth off.

When mortar cracks are small, a two-coat application of cement-sand grout, brushed vigorously into the mortar joints, will provide an effective and inexpensive waterproofing method. A typical mixture consists of equal parts of portland cement and dry sand, with one-quarter part of cement replaced by limestone flour, powdered flint, or fine hydrated lime. Thoroughly wet the joints before applying the grout.

If all mortar joints appear to be sound, but seepage continues, the entire wall surface must be coated with a waterproofing compound. There are several colorless waterproofing agents now on the market that do not change the color of the brick in any way.

When appearance is not a factor, a less expensive method of waterproofing consists of a single application of cement grout to the joints, followed with two coats of cement water paint.

If the wall is solid brick, dampness in the foundation and some low areas of the walls above grade caused by improper drainage or lack of waterproofing precautions is a common problem. Drain tile should be laid along the footings or a membrane waterproofing applied.

Water can enter the top of a brick veneer wall and dampen the plaster at the bottom. This can be remedied by drilling ¼-inch holes near the bottom of the brickwork along a horizontal joint in such a way that any water settling there will drain out.

Concrete Block and Stone Wall Problems The problems and solutions for brick walls are appropriate for block and stone walls. Most of the difficulties with stone walls require repointing the mortar joints.

Removing Stains from Masonry Surfaces

Almost any type of stain can be removed from masonry. When the source of the stain is unknown, the treatment may require some experimentation. Many chemicals can be applied to masonry without appreciable damage, but acids or chemicals causing violent reactions should be avoided.

For stains which have penetrated the surface, use a poultice or bandage. A poultice is made by mixing active chemicals and fine inert powder to a pasty consistency and is applied in a thick layer. Bandages are cotton batting or layers of cloth soaked in chemicals and pasted over the stain.

Paint Stains For fresh paint stains, apply a commercial paint remover or a solution of 2 pounds

Steps in repointing a stone wall. The stonework must be cleaned and all loose mortar removed before any repointing can be done. Also, the surface must be wetted down with water. (Use a brush to work the water into crevices.) Once the surface has been properly prepared, the new mortar can be applied.

trisodium phosphate in 1 gallon of water. Allow to stand, and remove the paint with a scraper and wire brush. Wash with clear water. For very old dried paint, organic solvents may not be effective. In this case, the paint must be removed by scrubbing with steel wool.

Efflorescence The term efflorescence, when applied to masonry, refers to the white powdery crust that sometimes accumulates on wall surfaces. It is composed of water-soluble salts, originally present in the masonry materials. It can frequently be removed by water applied with a stiff scrubbing brush. In those cases where this procedure does not remove all the efflorescence, the surface should be scrubbed with a solution of muriatic acid not stronger than 1 part commercial acid to 9 parts water by volume. It is important that you rinse with water both before and after acid baths. Give the surface a final washing with water containing approximately 5 percent household ammonia.

Ferrous Stains Iron stains are rust colored. The appearance of large stained areas can be improved by washing with a solution of 1 pound oxalic acid powder per gallon of water. After two or three hours, rinse with clear water, scrubbing with stiff brushes or brooms. Bad spots may be scrubbed with a second application.

For deeper stains, dissolve 1 part sodium citrate in 6 parts water. Mix thoroughly with an equal volume of glycerin. Mix part of this liquid with whiting to form a paste just stiff enough to adhere to the surface in a thick coat. Apply with a putty knife or trowel and allow it to dry for a few days. Replace with a new layer or soften by adding more liquid. Be sure to rinse it off once it has dried.

For deep and intense iron stains, use sodium hydrosulphite. Soak the surface with a solution of 1 part sodium citrate crystals and 6 parts water. Dip white cloth or cotton batting in this solution and put it over the stains for 10 or 15 minutes. Sprinkle horizontal surfaces with a thin layer of hydrosulphite crystals, moisten with water, and cover with a stiff paste of whiting and water. On a vertical surface, place whiting paste on a plasterer's trowel, sprinkle on a layer of hydrosulphite, moisten slightly, and apply to the stain.

Do not leave it on longer than an hour, or a black stain may develop. If the stain is not completely removed, repeat the operation with fresh materials. When the stain disappears, rinse the surface thoroughly with water.

Copper, Bronze, and Aluminum Stains Copper and bronze stains are usually green, but in some cases may be brown. Mix 1 part ammonium chloride (sal ammoniac) and 4 parts powdered talc, and stir in ammonia water to make a paste. Place this over the stain. Allow it to dry and brush it off. A stain of this kind requires several years to develop and may require several applications to remove. Aluminum chloride may be substituted for sal ammoniac.

Aluminum stain appears as a white deposit which can be removed by scrubbing with a 10 to 20 percent muriatic acid solution. On colored masonry, a weaker solution should be used.

Lubricating Oil Stains Lubricating oil penetrates some concrete readily. It should be mopped off immediately and the spot covered with Fuller's earth or other dry powdered material, such as hydrated lime, whiting, or dry portland cement. When the oil has remained for some time, saturate white flannel in a mixture of equal parts of acetone and amyl acetate and place over the stain. Cover with a slab of concrete or a pane of glass. If the stain is on a vertical surface, improvise some means of holding the cloth and its covering in place. Keep the cloth saturated until the stain is removed. If the solvent tends to spread the stain, a larger cloth should be used. Scrubbing with gasoline or benzine will often remove oil stains.

Another method of removing oil, grease, and soap spots is to place sawdust, sweeping compound, or fine wood shavings on such spots and soak overnight with a solution of 1 pound lye to a gallon of water. Dissolve the lye carefully so it will not spatter, and wear rubber gloves and goggles. Brush away the sawdust a day later and scrape the spot clean. Scrub the surface with a trisodium phosphate solution.

Plant Stains Occasionally an exterior masonry surface which is not exposed to sunlight remains in a constantly damp condition and may exhibit signs of plant growth, such as moss. Ammonium sulfamate is generally effective.

Smoke Stains Make a smooth, stiff paste of trichlorethylene and powdered talc and apply with a brush. Cover with a drinking glass or a pan to prevent rapid evaporation. Then rub it off. If a slight stain is left after several applications, wash thoroughly and use the method described for removing tobacco stains. Trichlorethylene fumes are harmful.

Soap and water applied with a stiff-bristle brush are frequently effective in removing soot and coal-smoke stains.

Tobacco Stains Dissolve 2 pounds of trisodium phosphate in 5 quarts of water. In a separate vessel, make a smooth, stiff paste of 12 ounces of chloride of lime in water. Pour the first solution into the paste and stir thoroughly. Make a stiff paste of this with powdered talc and apply in the same way as described for iron stains.

Decayed Wood Stains Under damp conditions, wood rots and causes a stain that is readily distinguished from most other stains by its dark color. The best treatment is the one recommended for smoke stains.

Mortar Stains These generally require the use of muriatic acid in a solution no stronger than 1 part acid to 9 parts water by volume. Before the acid solution is applied, the surface should be thoroughly soaked with clear water.

The acid solution should be applied with a long-handled fiber brush. Cover your clothing, hands, and arms to prevent burns. The solution should not be placed over an area greater than 15 to 20 square feet before the wall is again thoroughly washed down. Remove all traces of the acid before it attacks the mortar joints. All frames, trim, sills, or other installations adjacent to the masonry must be carefully protected against contact with the acid solution.

Index

Aggregate, 4, 84

Barbecue, 78
Brick, 37-44, 48-57, 69
  bonds, 40-41
  cleaning, 44
  cutting, 42
  grades, 37
  joint finishes, 43
  laying, 41
  mortar, 38, 41
  ordering, 39
  terminology, 40
  types, 37-38
Brick mortar, 38-41
  joints and pointing, 41
  preparing, 41
Brick patios, 69
Brick walls, 48-57
  cavity walls, 53
  footing, 48
  laying, 49-51
  openings, 51-53
  veneer, 54-55
  watertight, 55-56
Building regulations, 63

Cement, 3-4
Concrete, 3-16, 27-36, 57-59
  bull-floating, 11-12
  colored, 81
  curing, 15-16
  edging, 12
  floating, 13
  ingredients, 3-4
  jointing, 12-13
  mixing, 5-8
  pouring, 10-11
  premixed, 8
  preparation, 9
  ready-mixed, 8
  repair, 87-95
  strikeoff, 11
  troweling, 13-14
Concrete block, 27-36, 57-59
  cutting, 34
  estimating materials, 32

laying, 33-34
  modular planning, 27
  mortar, 29-32
  patching, 35, 89-90
  sizes, 27
  walls, 21, 57-59
Concrete block walls, 21, 57-59
  control joints, 58
  finishing, 59
  foundation, 57
  lintels, 59
  structural, 58
Concrete floors, 25
Concrete patios, 68
Concrete slabs, 23-25
Crawl space, 22-23
Curbs, 68

Dampproofing, 55-56, 90-93
Driveways, 62-68
  formwork, 65-66
  planning, 62-63
  pouring, 67-68
  preparation, 64

Finishes, 81-86
  dusted on, 81
  dyes, 83
  exposed aggregate, 84
  mineral oxide, 81
  painted, 81-82
  patterns, 85
  sparkling, 86
  stains, 83
  textured, 85
Fireplaces, 78-80
Flashing, 56
Floors, 25
Footings, 17-20, 48
  forms, 20
Foundations, 17-25, 48

Garden walls, 72

Hand mixing, 7

Lintels, 52-53, 59

Machine mixing, 6-7
Maintenance, 87-95
Masonry tools, 28-29
Mortar, 29-32
  mixing, 31
Mortarless block, 35-36

Patios, 63, 69-70
  formwork, 65-66
  planning, 63
  preparation, 64
Precast slabs, 70-71

Repair, 87-95
  chips, 89
  cracks, 87
  dampproofing, 90, 93
  holes, 89-90
  stains, 93-95
Retaining walls, 76-78

Sidewalks, 63, 69-71
  formwork, 65-66
  planning, 63-64
  preparation, 64
Steps and porches, 72-75
Stone masonry, 44-47
  cleaning, 47
  cutting, 46-47
  joints and pointing, 46
  laying, 46
  materials, 45
  mortar, 45
  walls, 59-61
Stone patios, 69-70
Stone planters, 75
Stone walls, 59-61
  building, 61
  structural, 59-60
  veneer, 60

Veneer, 54-55, 60

Walls, 21, 48, 57, 61
  block, 57-59
  brick, 48
  concrete, 21
  stone, 59-61